ILEIA READINGS IN SUSTAINABLE AGRICULTURE

Linking with Farmers

Networking for Low-External-Input and Sustainable Agriculture

Edited by Carine Alders, Bertus Haverkort and
Laurens van Veldhuizen

INTERMEDIATE TECHNOLOGY PUBLICATIONS 1993

Practical Action Publishing Ltd
27a Albert Street, Rugby, CV21 2SG, Warwickshire, UK
www.practicalactionpublishing.org

© ILEIA, 1993

First published 1993\Digitised 2008

ISBN 13 Paperback: 9781853392108
ISBN Library Ebook: 9781780445328
Book DOI: http://dx.doi.org/10.3362/9781780445328

All rights reserved. No part of this publication may be reprinted or reproduced or utilized in any form or by any electronic, mechanical, or other means, now known or hereafter invented, including photocopying and recording, or in any information storage or retrieval system, without the written permission of the publishers.

A catalogue record for this book is available from the British Library.

The authors, contributors and/or editors have asserted their rights under the Copyright Designs and Patents Act 1988 to be identified as authors of their respective contributions.

Since 1974, Practical Action Publishing has published and disseminated books and information in support of international development work throughout the world. Practical Action Publishing is a trading name of Practical Action Publishing Ltd (Company Reg. No. 1159018), the wholly owned publishing company of Practical Action. Practical Action Publishing trades only in support of its parent charity objectives and any profits are covenanted back to Practical Action (Charity Reg. No. 247257, Group VAT Registration No. 880 9924 76).

Typeset by J&L Composition Ltd, Filey, North Yorkshire, UK

Contents

	Page
Preface	v
Acknowledgements	vi

Part 1: Networking for Low-external-input and Sustainable Agriculture — 1

Bertus Haverkort, Laurens van Veldhuizen and Carine Alders: — 3
Networking for low-external-input and sustainable agriculture

David Korten: — 25
Strategic networking: From community projects to global transformation

Part 2: Farmers' Networks — 35

Bertus Haverkort: — 37
Farmers' networks: Key to sustainable agriculture

Eric Holt-Giménez and Orlando Cruz Mora: — 51
Farmer to farmer: The Ometepe Project, Nicaragua

Wim Hiemstra, Fokke Benedictus, René de Bruin and Pieter de Jong: — 67
Farmers search for new ways of cooperating: Networking in the Friesian woodlands of the Netherlands

Ron Kroese and Cornelia Butler Flora: — 81
Building a grass roots institution to implement low-external-input and sustainable agriculture: The stewardship farming experience

Stephan Rist: — 93
Supporting indigenous knowledge for sustainable rural development in Bolivia: The case of AGRUCO

Janice Jiggins: — 109
Networking with women farmers

Frances Kinnon: — 117
Farmer networking at international level

Part 3: NGO Networks — 129

Paul Engel: — 131
Daring to share: Networking among non-government organizations

Enrique Kolmans: — 151
Networking for sustainable agriculture in Peru: Experiences of the Red de Agricultura Ecológica

Malex Alebikiya: 159
The Association of Church Development Projects in Northern Ghana
Jorge Manrique, Juan A. Palao and Mourik Bueno de Mesquita: 167
Andeans unite: The birth and growth of the Andean Council of Ecological Management
Oswald Quintal and Gandhimathi: 177
Networking among resource-poor farmers in South India

Part 4: Research Networks 185
Donald L. Plucknett, Nigel J.H. Smith and Selcuk Ozgediz: 187
Networking in international agricultural research
Paul Starkey: 199
Animal traction networks in Africa: Lessons and implications
Gordon Prain, Virginia N. Sandoval and Robert E. Rhoades: 215
Networking for low-external-input and sustainable agriculture: The case of UPWARD

Part 5: Support Organizations 229
Terry Smutylo and Saidou Koala: 231
Research networks: Evolution and evaluation from a donor's perspective
Irene Täuber, Almut Hahn and Claudia Heid: 249
Support to networking in Africa and Latin America: The role of AGRECOL
Wim Hiemstra and Carine Alders: 259
Supporting regional networks: The experiences of ILEIA
Olivia Graham: 271
Networking as a development activity: The Arid Lands Information Network

Further Reading 281

List of Networks 287

Preface

Pascal Badjagou, of the Réseau de Développement d'Agriculture Durable (REDAD), in Benin, describes his experiences in starting a network thus:

> More often than not, because of lack of communication, people tend to look far away for something that is in fact close by, without their being aware of it. For instance, we learnt about the existence of agroforestry from books and magazines, but we were overjoyed, when making our initial contacts to form a network on it, to discover the RAMR project (Recherche Appliquée en Milieu Réel), which practises agroforestry in Mono Province, only 150 kilometers from our community!

He and his colleagues are typical of the 'prime movers' whose drive and enthusiasm are seminal in the launching of a new network. But will their efforts meet with success? And if they do, how will that success be measured? Will the network last? Will it remain informal or become institutionalized?

These are some of the questions addressed by this book, which describes the experiences of researchers, development workers and farmers worldwide, as they attempt to build working relationships through networking. Because it brings people together in a common endeavour, networking can play a major role in strengthening the impact of their efforts. Sharing experiences, opinions and information strengthens self-confidence and makes participants better equipped to manage a complex and rapidly changing environment. Networking is especially important in the development of low-external-input and sustainable agriculture, which has received little support from conventional research and development institutes in the past.

We hope that, by sharing their experiences, the contributors to this volume will inspire readers to new efforts and initiatives in networking for the future development of low-external-input and sustainable agriculture.

Carine Alders, Bertus Haverkort and Laurens van Veldhuizen (editors)

ILEIA
P.O. Box 64
3830 AB Leusden
Netherlands

Acknowledgements

The foundations for this book were laid during an International Workshop on Networking for Low-external-input and Sustainable Agriculture which took place in Silang, the Philippines, in March 1992. Jointly organized by the International Institute of Rural Reconstruction (IIRR), World Neighbors and ILEIA, this workshop was attended by some 45 representatives of different networks from 20 countries in North and South America, Europe, Africa and Asia.

The contributors to this book all lead busy lives. They fulfil a position in their own organization, but at the same time they put energy into networking—energy that is often unbudgeted in terms of either time or financial resources. On top of these already heavy demands they have found time, at our request, to write down their experiences and to answer the many questions we had. This book could not have been published without their enthusiasm and commitment.

Publishing a book is a huge job. We are lucky to have had support from many dedicated people. We extend our special thanks to Julian Gonsalves, Scott Killough and all other staff of IIRR, who truly hosted the workshop and helped us to shape our thoughts about networking, and to Larry Fisher of World Neighbors, who time and again translated our vague, abstract ideas into plain and practical language. Both he and Ilyia Moeliono contributed considerably to the editorial. Simon Chater, of Hawson Farm, Devon U.K., edited the text, polishing up our English (or what passed for English) and providing useful criticism in so doing. With the invaluable assistance of Christel Blank, he also made the book camera-ready for the publisher. Finally, we are very grateful to Puck Sluijs, for making the line drawings, to Studio Dryia Media, for their eye-opening artistic work, and to Lila Felipie, Marika van den Brom, Lyanne Alons, Ellen Radstake, Marian de Boer and Yvonne van Toren for their reliable administrative support.

PART ONE

Networking for Low-external-input and Sustainable Agriculture

Networking for Low-external-input and Sustainable Agriculture

Bertus Haverkort, Laurens van Veldhuizen and Carine Alders

Introduction

Networking is a powerful and cost-effective way of sharing information and achieving various other goals that individuals cannot achieve alone. During and after an international workshop which took place in Silang, Cavite in the Philippines in March 1992, participants exchanged their experiences of and through networking. The workshop was attended by 45 representatives of different networks from 20 countries of North and South America, Europe, Africa and Asia, and was jointly organized by the International Institute of Rural Reconstruction (IIRR), World Neighbors and ILEIA. This paper combines observations made at the workshop with the other networking experiences described in this book.

To understand and appreciate the role networking can play in low-external-input and sustainable agriculture (LEISA), let us first look at the main characteristics of LEISA and the related notion of participatory technology development (PTD).

Reijntjes et al (1992) give the following definition of LEISA:

> Low-external-input and sustainable agriculture refers to those forms of agriculture that: seek to optimize the use of locally available resources by combining the different components of the farm system, i.e. plants, animals, soil, water, climate and people, so that they complement each other and have the greatest possible synergetic effects; and seek ways of using external inputs only to the extent that they are needed to provide elements that are deficient in the ecosystem and to enhance available biological, physical and human resources. In using external inputs, attention is given mainly to maximum recycling and minimum detrimental impact on the environment.
>
> LEISA does not aim at maximum production over a short duration but rather at stable and adequate production over the long term. LEISA seeks to maintain and, where possible, enhance natural resources and make maximum use of natural processes. Where part of the production is marketed, opportunities are sought to regain the nutrients brought to the market. At farm, regional and national level, LEISA implies the need for close monitoring and careful management of the flows of nutrients, water and energy so as to achieve a balance at a high level of production. Management principles include harvesting water and nutrients in the watershed, recycling nutrients within the farm, managing nutrients in the flow from farm to consumer and back again, using

aquifer water judiciously, and using renewable sources of energy. As these flows are not confined by farm boundaries, LEISA requires management not only at farm level but also at district, regional, national and even international levels. At each level, technologies should be sought to make the flow cycle as short as possible and to balance the flows.

LEISA incorporates the best components of indigenous farmers' knowledge and practices, ecologically sound agriculture developed elsewhere, conventional science and new approaches in science (e.g. the systems approach, agro-ecology, biotechnology). Thus far, conventional science has served mainly high-external-input agriculture, but the contributions it could make to LEISA should be explored to the full. LEISA practices must be developed within each ecological and socio-economic system. The specific strategies and techniques will vary accordingly and will be innumerable.

Developing the New Agriculture: A Role for Farmers and Researchers

In this search for eco-specific and socio-economically adapted farming techniques, farmers and external professionals can both play a role. One form of the cooperation between them is known as participatory technology development (PTD) (ILEIA, 1989).

Conventional research and technology development policies have been criticized for failing to significantly improve low-external-input agricultural systems, having focused mainly on irrigated agriculture and export crops (Chambers and Jiggins, 1986; OTA, 1988; Arbab and Prager, 1991). The technologies developed have often proved inappropriate for resource-poor farmers and pastoralists. Many technical interventions have also been ill suited to women's needs and priorities. Furthermore, the conventional approach has led to technologies that demand the use of high levels of external inputs, endangering the environment and undermining the sustainability of agriculture.

In response to these criticisms, major changes in the approach to both research and development are now taking place. Scientists are seeking ways of reducing the levels of external inputs used. The international agricultural research centres, which have long worked on increasing the sustainability as well as the productivity of agriculture through a commodity-focused approach, have recently placed renewed emphasis on the management of natural resources, developing an ecoregional perspective to complement their global commodity research. The study of multiple cropping, the roles of livestock, multi-purpose trees and agroforestry systems have all increased dramatically in recent years. It is now generally accepted that rainfed agriculture in more fragile environments cannot be modernized using the same approaches as those of the Green Revolution. In these areas, relatively low levels of external inputs need to be combined with maximum utilization of locally available resources and natural processes. Some notable progress is now apparent in the improvement of crops grown by the resource-poor farmers who live in rainfed areas, including cassava, millet, maize, cowpea and pigeonpea. The Food and Agriculture Organization (FAO) of the United Nations now advocates what it calls sustainable agriculture and rural development (SARD), which emphasizes both integrated pest management (IPM) and integrated plant nutrition systems (IPNS). Meanwhile, the scientists have gone a stage further, replacing the concept of integrated pest management with that of plant health, which emphasizes prevention rather than cure and the avoidance of chemicals altogether wherever possible. Technologies in plant health have made particularly rapid progress in recent years, with the development of genetic resistance and of biological control helping governments and farmers to cut down radically on the importation and use of pesticides in certain cases. In addition, there has been widespread recognition of the importance of including farmers and other land users in the process of technology development. Social and economic factors, including the importance of women's contributions to agriculture, are now far better understood than they were when criticisms of the Green Revolution were first voiced more than 20 years ago.

PTD builds on and contributes to these changes in approach, blending the experiences of practitioners of farming systems approaches to research with those of NGO staff engaged in community-based rural development. It differs from conventional approaches to research and extension in several important respects:

- It gives special emphasis to the indigenous knowledge of farmers, seeking to use this as a basis for technology development.
- It seeks to enhance farmers' capacities to develop technology and to secure their inputs at all phases of the research cycle, particularly the crucial technology design stage.
- It seeks to involve all members of the community in research and development activities, especially those traditionally disadvantaged by conventional research and development approaches (such as women).
- PTD aims at developing a variety of options for further experimentation rather than a fixed package to be implemented as it is.
- Methods used are generally rapid, low-cost and locally manageable.
- PTD does not claim to be a fixed and complete model, but an approach that can and should be adapted according to experience and local circumstances.

To be effective, PTD activities depend on collaboration among farmers, fieldworkers, researchers and others. New patterns of interaction and cooperation between these actors are therefore needed. Networking has a key role to play here.

The Importance of Networking

Networks can play an important role in bringing about a more participatory approach to agricultural research and development. A growing number of farmers' groups, NGOs, research and development organizations and donor agencies are becoming interested in the development of LEISA. Yet pioneers in this field realize that its application is not spreading as fast as it could and should. This is because of two crucial bottlenecks. First, many organizations dedicated to agricultural development are small. They work in relative isolation and have difficulties in finding qualified staff. They silently build up a wealth of experiences which do not reach other organizations working in the field. Second, the socio-economic and policy environment in which both farmers and development organizations must operate is often not conducive to sustainable agriculture. Networks also have a role to play in addressing this second bottleneck. As removing these bottlenecks is essential to the further development of LEISA, they are further discussed below.

Strengthening development support organizations

The tasks faced by development support organizations are many and complex. They need to understand their own place in the political game, take stands and voice their opinions in the most appropriate way. In the field of agriculture they need to be aware of new technologies and practices that may suit the needs of the people they serve. Above all, they need skills in participatory methods of working with farmers. Further, they must develop links with government agencies to obtain support for their field programmes. In all their activities they are subject to the vicissitudes of national and international political life. Heavily engaged in running their own programmes, few development organizations have sufficient time to keep abreast of new ideas and options. Recent experiences from a nearby organization may go unnoticed, let alone those of organizations in other parts of the world.

Great efforts are needed to improve the effectiveness of these organizations. Development strategies need to be designed, human resources need to be developed, documentation and monitoring and evaluation need improvement, and links with sources of new information need to be established. Single organizations acting alone find it extremely difficult to do these things. They need to collaborate with others.

Changing the policy environment

Conventional agricultural development policies, with their concepts, criteria, procedures and institutional structures, have been designed mainly on the premise that agricultural intensification requires high levels of external inputs. Until recently, most countries (developed as well as developing) adopted policies which, through subsidies, encouraged monocropping and the use of chemicals. Price interventions supported specialization and the switch to commercial crops. The detrimental environmental effects of such policies were generally externalized, so that the rest of society paid. These policies neglected or undervalued local knowledge and generally had a male bias. Policies were generally applied nationwide, without taking into account cultural, economic and biophysical diversity.

At a global level, both producer and consumer prices of agricultural products are greatly influenced by international market relationships. Consumer prices are lowered by the subsidized exports of Western countries; exports from developing countries are restricted by various forms of trade barrier imposed by the industrialized North, and sometimes even by export taxes in the country of origin.

Policy support to agricultural development needs to be modified to make the transition from unsustainable forms of agriculture to more sustainable forms. Changes in policies should include more support for low-external-input approaches to technology development, the introduction of pricing policies that favour the use of local resources, participatory and decentralized planning methods, and legislation for environmental protection. Experiences in these areas are few and far between as yet—but they do exist. In recent years structural adjustment programmes in many countries have led to a new economic environment, in which external inputs have largely lost their comparative advantage over the use of local resources. Experiences with LEISA and PTD are being better documented and are gaining recognition. International agreements are leading to new restrictions on the use of chemicals.

Shifts in agricultural policies will not come automatically. The status quo in both the North and the South favours existing elites, for whom it provides jobs, security, income, status and prestige. For these elites, any pressure for change that favours the poor at their expense will be seen as a threat. A major innovative role therefore has to be played by farmers' groups, NGOs and research organizations active in LEISA. In this respect too, networking will be important.

Networking and empowerment

Networking can empower network members who belong to an otherwise isolated or marginalized social group. Women's networks function as a mechanism to counter male domination (see Jiggins, p.109). Some networks give farmers back their self-confidence by strengthening or reviving indigenous knowledge systems (see Haverkort, p.37, and Rist, p.93). Others validate farming styles that challenge the progressive farm model espoused by government (see Hiemstra et al, p.67). Networking may strengthen Southern farmers' organizations in their confrontation with Northern-dominated market and technology development systems (see Kinnon, p.117), or give a voice to farmers who are dissatisfied with government policies (see Hiemstra et al; Holt-Giménez and Cruz Mora, p.51; or Compton and Joseffson, cited in Haverkort). Finally, networks strengthen environmental groups in their attempts to prevent environmentally destructive development schemes (see Korten, p.25).

North-South relationships

Many research organizations and NGOs of the South have been established by and/or receive considerable funding from donor agencies based in the North. (Some of these NGOs have a specific ideological or religious mandate.) Both project-based and core funding may be involved. This means that their work plans and budgets have to be approved by people in the North. Such dependence may introduce a Northern bias in the choice of priorities and criteria for planning and evaluation.

The lack of exchange of experiences between NGOs in a specific region may partly be caused by the fact that, in their external relationships, they focus on organizations with the same ideological flavour rather than on those working in a similar development context. The Northern support organization sometimes plays a part in fostering this tendency. Establishing regional networks is an important mechanism for reducing donor influence of this kind. In this respect the support now being provided to the development of oecumenical networks by donor agencies based in the North is especially welcome. The cases of the Association of Church Development Projects (ACDEP) and the Ecological Association for Sustainable Agriculture and Rural Development (ECASARD) in Ghana (see Alebikiya, p.159) are examples.

Donor organizations funding research are heavily biased towards Western technological and scientific tradition. This tradition is based on reductionist paradigms and fosters highly specialized research efforts, focusing on single aspects of a problem. This form of research has been developed in the North to serve the needs of the market economy. The fact that Southern research institutes are oriented towards this Northern tradition is in part a result of past North-South relationships. Many national agricultural research systems are still organized along the lines of former colonial institutions. Senior positions are often held by Northern scientists. Many national researchers have received their training in the North, and scientific publications are generally produced by Northern organizations. The International Development Research Centre (IDRC) and the Swedish Agency for Research Cooperation with Developing Countries (SAREC) are to be commended for their role in supporting trans-disciplinary research networks.

Much homework remains to be done by funding agencies. In the Netherlands, for example, there are four different ideology-based funding agencies, all active in NGO support. The government agency, the Directorate General for International Cooperation (DGIS), is still influenced by specialists who lack a holistic approach. At the international level, North-South relationships frequently still mirror history: francophone, anglophone and Arabic speaking countries tend to have their own historically determined links with specific 'mother' countries; the USA has Latin America as its 'backyard', while Europe 'has' Africa and Japan and Australia 'have' Asia. Increased networking among Southern NGOs and researchers will lead to better internal coordination and improved 'donor management' in the interests of the South. As such it can only contribute to the further emancipation and empowerment of Southern organizations.

A Closer Look at Networking

Definition
There are several definitions of networking, each depending on a particular perspective. We use the following definition:

A network is any group of individuals and/or organizations who, on a voluntary basis, exchange information or goods or implement joint activities and who organize themselves for that purpose in such a way that individual autonomy remains intact.

In addition to this definition, the meaning of networking may become clearer by listing some of its most important characteristics:
- Members take part on a voluntary basis; networking assumes the willingness to share information and other resources in an environment of mutual trust and respect.
- Members carry out joint activities that cannot easily be performed alone.
- Members' individual autonomy remains intact.
- Networks can have many different forms and use different procedures depending on the specific situation. There is therefore considerable diversity in networking experiences.
- The network's structure is often 'light' and not very formal.

Networking has long been recognized by research institutes and funding agencies as an important way of improving the effectiveness of agricultural research (Plucknett, 1990). Similarly, farmers generally take part in several local farmers' networks and can derive great benefit from them (Box, 1989; Compton and Joseffson, 1993). NGOs have tended to value their independence very highly, but are presently linking up more and more with their peer groups. However, no systematic assessment of the experiences of these different groups in networking for LEISA has yet been made.

Typology

There are several types of network and several different criteria on which a typology can be based. Figure 1, designed during the workshop in the Philippines, gives a typology on the basis of the professional background of members and the activities involved.

Figure 1 Typology of networks according to membership and activities

	Activities:					
	Information exchange	Materials exchange	Training	Marketing	Awareness raising	Policy dialogues
Membership:						
Farmers' organizations						
Non-government organizations						
Researchers and extensionists						
Various categories						

Another typology can be made according to the flow of information and internal organization of the network. Depending on the degree of centralization and hierarchy, the types of network shown in Figure 2 can be distinguished.

Figure 2 Typology of networks according to flow of information

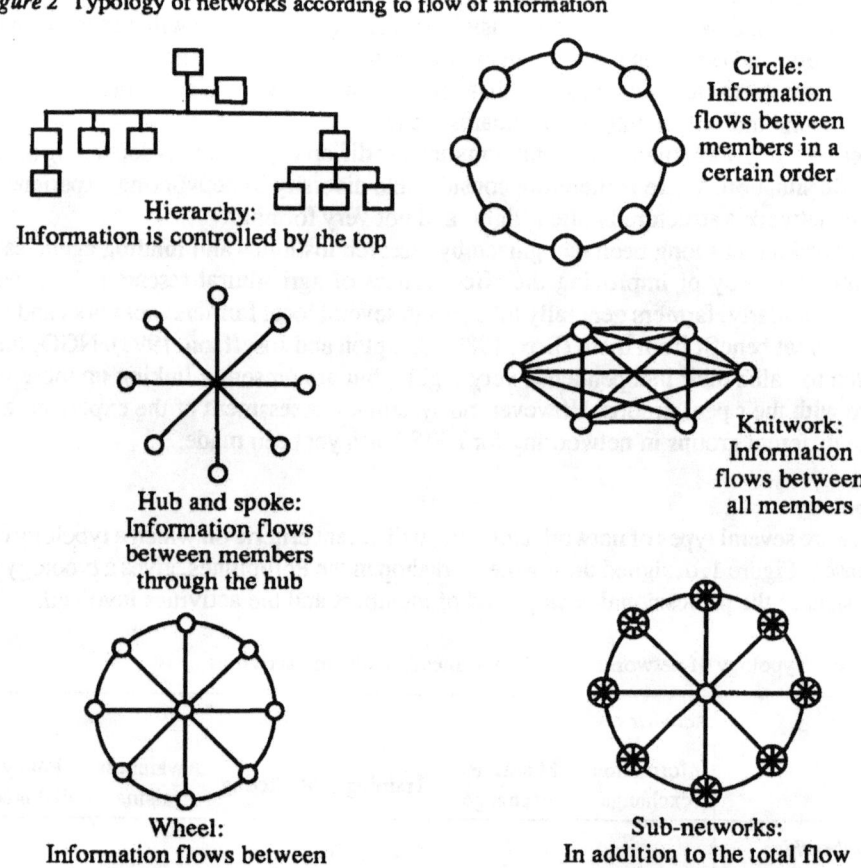

Source: Haverkort and Ducommun, 1991

Plucknett et al (p.187) give a typology based on operational styles. Further typologies could be based on scope of geographical coverage (local, national, regional, international), or on subject matter focus (e.g. networks on integrated pest management, on soil fertility, etc).

Analyzing the different forms and types of network that have evolved or have been chosen by members reveals certain patterns. Table 1 presents a preliminary analysis of this kind. This may help people define the most appropriate networking model for their situation. The paper by Manrique et al (p.167) gives a similar table, based on the experiences of the Andean Council of Ecological Management (CAME).

Table 1 Advantages and disadvantages of different network types

Type	Advantages	Disadvantages
Local level	Allows face-to-face contact	Limited scope and means
	Eco-specific	
	Informal	
(Supra)national level	Represents large numbers of people	Requires formalization and core funding
	Allows stronger policy voice	Limited interaction between members
	Acquires more resources for large tasks	
Specialized subject	Well focused	Too narrow a focus
General focus on LEISA	Holistic	Too dispersed
Horizontal membership	Deeper contacts	
Vertical membership	Allows contacts between levels	
Centralized organization	Executive power	Alienation from grass roots
	Easy for donors to deal with	
Decentralized organization	Democratic	Difficult to maintain cohesion
	Commitment	

The Emergence and Evolution of Networks

The inventory carried out during the workshop held in the Philippines showed that many networks to enhance LEISA have been established in recent years. Networks at continental level had been formed in Europe (the European Network on Low-external-input and Sustainable Agriculture, EULEISA), Latin America (the Latin American Consortium for the Development of Ecological Agriculture, CLADES) and South-East Asia (the South-East Asian Sustainable Agriculture Network, SEASAN). National and sub-national networks are already plentiful—and many more are in the process of being formed.

Preconditions for networking

Not all experiences with networking have been positive. Many initiatives have failed. Recent experiences suggest that the following questions need to be considered before starting a network:

- Is there a common vision and a set of common goals among potential members?
- Do potential members face common problems and constraints?
- Are potential members aware of these problems and constraints and of their important influence on their work?
- Do potential members, especially NGOs, have sufficient managerial and organizational skills?
- Are there relevant results/experiences that can be shared?
- Do potential members have a good idea of what a network is and what it could mean to them?
- Are they prepared to spend the necessary time and energy in sharing and networking at the expense of time spent with members of their own programmes?
- Is there an atmosphere of openness among potential members which allows them to admit mistakes and learn from them?
- Can the coordination of a network be ensured, especially during its first phase of emergence?
- Can the necessary financial resources for network activities be mobilized from network members? If not, is there a chance of continuous donor funding?
- Is there enough commitment to overcome the problems of the network's establishment phase?

Development stages of a network

Given the great diversity of networks, it is obvious that there is no blueprint for their development. In reality, networks usually evolve slowly and follow a development path dictated by their own internal logic. Farmers' networks tend to be less formal than networks among other groups.

However, the analysis of different experiences enables some stages to be identified which appear relevant for most networks. These are:

Preparation. During this stage initiators identify a topic of common concern, formulate the idea of a network, and assess the interest of potential members. For any network it is important that this stage be based on the perceived needs of the potential members. These can be assessed through a preliminary survey. In informal networks these activities take place spontaneously.

In formal networks, however, the initiator plays a catalytic role which requires time, thought and financial resources. In research networks the host organizations generally provide these resources. In return they are able to influence the network in a direction they deem important. Finding financial resources is not easy during the preparation stage, because it is not yet certain whether the network will take off. A survey may indicate that there is not enough interest, or that networking is not the right organizational form with which to tackle the problem. In some cases, such as the Tamil Nadu network of NGOs described by Quintal and Gandhimathi (see p.177), a supporting organization such as ILEIA can provide seed money to meet preparation costs. The possible roles of support/donor organizations during this stage are summarized in Box 1. The issues raised are based on the experience of several recently established networks in India and Africa.

Establishment. During this stage the decision is taken to form a network, typically at a founding meeting of all potential members. The members also determine the mechanisms and structure for information exchange and collaboration. In some cases networks draw up formal rules and regulations, with a central committee and a well defined membership, and then organize funding. In other cases, establishment occurs informally, with the mechanisms and structure evolving according to need. Whether formally or informally, it is essential that members participate in planning activities at this stage. The relationship of the network with the initiators, the host institution and the donor has to be defined. The experiences of IIRR indicate that when a formal network is established it is important to keep overhead costs low and network structures simple. The maintenance of a secretariat and office equipment can be both costly and cost-ineffective unless real collaborative activities develop in the field. Formal networks can easily become yet another bureaucratic layer inhibiting rather than promoting the exchange of information and cooperation. As Grimaldo Rengifo of the Andean Project for Indigenous Technology (PRATEC) said, 'Structuring life can freeze life' (personal communication).

Operations. After establishment the network gradually becomes fully operational. It adapts according to environmental change and internal dynamics. The latter will be greatly influenced by personalities. A clear identification of the network's goals, structure and procedures and some training in network management will help guide the

> **Box 1. Possible roles of support/donor organizations during network formation**
>
> - First, development support organizations can link with emerging initiatives. In certain cases they can initiate by approaching leading NGOs and inviting them to explore the possibility of forming a network. In this way they can play a catalytic role in breaking down existing patterns of non-collaboration.
> - An ad-hoc committee or core group can then be formed, consisting of representatives of different NGOs. To allow the committee to do its preparatory work, the support organization can make some seed money available to cover travel and communication costs.
> - The committee conducts a survey of the need for a network and of the available experience and expertise to be shared through it. On the basis of this survey, a register of members is made and the feasibility of the network becomes clear. A draft statement of intent on the network's purpose is formulated and presented to potential members. At this stage the support organization can link the members of the committee or core group to other emerging networks, for example by informing them of examples of registers from elsewhere in the world, or by organizing exchange visits.
> - A constituting meeting can then be organized, to which all potential members are invited. The meeting should agree on the intention, objectives, structure and activities of the network. The role of the support organization may be to fund and/or to host the meeting.
> - Once the network has been formed, its finances need to be addressed. Networks should mobilize funds from their own resources to the extent possible. If these are insufficient a funding proposal can be developed and presented to donor agencies. A support organization may provide the necessary liaison at this point.
> - Technical support to emerging networks should preferably be provided by networks with a similar background. Such a South-South support system would benefit from a register of similar existing networks. A supporting or donor agency may possess such a register or may make its development possible.
> - Donor agencies should give priority to funding networks dedicated to LEISA on a national or sub-national basis as a cost-effective way of promoting sustainable agriculture.

network through this stage, which will probably be punctuated by various crises. Some redesign of the network's structure and management may be necessary to allow the membership to become more fully active. This stage is, of course, the most important one in terms of programme implementation. Several difficult issues may arise in the course of implementation. These are generally not related to how to make a newsletter or organize a workshop, but more often to internal cooperation and the management of resources. These issues will be discussed in more detail later.

Dissolution. According to Plucknett et al (1990) most networks should eventually disband, allowing members to regroup and confront new problems. Networks set up to tackle specific problems can dissolve once the task has been accomplished or the problem proves intractable. Networks should therefore consider themselves as temporary organizations. In practice, however, relatively few of the international research networks formed in recent decades have ceased to operate. Instead of dissolving a network, it may make more sense to transform it to address new issues, or to merge it with other networks. The following questions have to be asked before a network is disbanded or transformed: has the task been completed? If not, what parts of the task remain to be done? Do opportunities for fruitful collaboration still exist in some areas? What changes are necessary to meet the needs of the new situation? If the network is to be phased out, what final concerns should it address, and how will its residual responsibilities be met after dissolution?

Network activities

Experiences show that networks pursue a great variety of activities to achieve their objectives. However, some activities are common to almost all networks. Managing some kind of documentation or information centre or service is one of these. Here, information is pooled, processed and made accessible to all members. Information exchange is often promoted through a newsletter, varying from a two-page irregular bulletin to a regular, sometimes glossy, magazine. Further common instruments in networking are workshops and exchange visits. It goes beyond the scope of this book to describe these activities in detail and to say how they should be planned and implemented so as to become effective networking tools. The book *Information exchange networking for agricultural development: A review of concepts and practices*, by J. Nelson and J. Farrington, to be published in 1993 by the Technical Centre for Agricultural and Rural Cooperation (CTA) in English and French, contains valuable information and guidelines on networking and network management.

Managing Networks

Issues

Management issues in networking may relate, for example, to the way information is handled, the issue of leadership, the handling of differences between members, and the maintenance of members' commitment. All networks consider rotation of leadership important to avoid monopolization. The internal processes of management should be evaluated periodically, preferably with the help of outsiders. When there is a full-time secretariat, a separate steering committee should be appointed to make policy decisions and to avoid alienation of the secretariat from the members. The main role of the network's technical coordinators, often responsible for operations at regional level, is to ensure that members remain committed to the network. Managers and coordinators need to master the art of encouraging others, creating an atmosphere of mutual confidence and enthusiasm. Differences in technical approaches between members should be handled by accepting both points of view as legitimate rather than by forcing either or both of them to reach a uniform solution. Differences between members over network activities and procedures should be managed by delegating decisions to sub-committees and by encouraging regional activities. Managers must strike a balance between taking sufficient control to make the network more than the sum of its constituent parts, and allowing sufficient flexibility so that the individual members feel the relevance of activities for their own situation.

Many networks probably need staff training in the management of information. Research networks give special importance to improving communication by bridging language barriers and breaking down isolation. Care should be taken that members who have limited access to electronic media and to fax machines are not excluded from the flow of information.

The international support networks have the task of monitoring the different flows of information and suggesting initiatives for site-specific or topic-specific workshops, newsletters, documentation systems, review meetings, lobbying activities, and so on. They have to assess the quality of the information being exchanged and to take decisions accordingly. They may sometimes have to disappoint individuals by, for example, deciding not to publish a contribution in a global newsletter but referring it back to the regional network. They may ask probing questions to stimulate the author to develop the contribution further. Support organizations therefore have an important role in promoting the quality, not just the quantity, of networking.

Monitoring and evaluation

The effects of networking need to be monitored and evaluated continuously. Almost all the cases described in this book stress this point. Networking is a relatively new activity for most professionals in the field of sustainable agriculture. The art of networking can best be learned by accumulating learning experiences. So far, most of the experiences have been based on trial and error, but increasingly in the future the stock of knowledge should provide the basis for formulating practical guidelines. Yet it should also be borne in mind that networking is not an activity to be carried out on the basis of blueprints; each network will develop its own optimal form through its own experience.

Monitoring and evaluation are essential for making learning experiences explicit so that these can be shared with peer groups and donor agencies, and the efficiency and effectiveness of networking improved. Critical questions include: do members perceive benefits from the network? How do the structure and activities of the network relate to its objectives? How does management relate to members and how participatory are planning activities? Do networking activities improve the performance of individual members and their organizations?

After reviewing recent networking experiences among NGOs, Engel (p.131) develops a framework for structuring evaluation questions such as the above according to the objectives or function of the network. He uses a distinction between the efficacy, efficiency and effectiveness of networks and suggests indicators for network performance and impact.

Facing the Problems

Several problems in operating networks were identified by participants at the workshop in the Philippines. How best to structure and manage networks? How to acquire the necessary financial and human resources? How to monitor network performance and evaluate impact? Various ways of overcoming these problems were described by World Neighbors in its newsletter *World Neighbors in Action* (Fisher and Moeliono, 1992), from which the following section borrows heavily.

Unclear objectives

Some networks do not seem to have a clear purpose. Either the organizers never established the objectives or they did not let members know exactly what they were. If this happens, members have difficulty in knowing what to do. They quickly become confused or lose interest. Without a clear vision or strategy, it is difficult to make the right decisions, and the organizers may end up making decisions that fit their own personal interests.

The impact of any network can only be judged in relation to the network's goals, which therefore need to be set early on. These goals will determine the types of people and organization that will join and the kinds of activity that the network will carry out. The objectives should be: (a) clear, (b) decided upon by as many of the participants as possible, and (c) reviewed regularly to see if the network is having the desired impact and to make sure its goals are still relevant.

Domination

In networks involving institutions of different types, there will be different ideas about what the network should do. This can be positive, providing participants with an opportunity to challenge one another and so to stimulate each other's thinking until common ground is established and differences are worked out. However, if differences become too serious, competing groups may separate and the network will become paralyzed.

Sometimes a network is started by a few individuals or organizations who insist that those who join must share their views exactly. In other cases a small group may gradually begin to dominate decision-making and activities. If people bring a political agenda or personal or institutional conflicts to the network's meetings, small groups may begin to dominate while others are kept out of decision-making, being labelled as too academic, too far left, too far right and so on.

For the greatest impact, a network should include a variety of people and institutions that can relate to its objectives. Coalitions of many institutions and groups will stimulate debate and encourage collaboration, both of which are major reasons for networking. Steps need to be taken early to make sure most decisions and activities are made and carried out democratically.

While it may be difficult to stop fighting between groups, no harm is done by reminding them that the purpose of the network is to work with rather than to exclude one another.

Tight central control

The domination of a network by a small group usually means that decisions will be made by the select few. This can make it difficult for the network to change, grow or accept new ideas. If setting up a tight central bureaucracy (in the form of a secretariat, organizing board or steering committee) puts still more control in the hands of the few, the centre can become a tool for control rather than coordination.

While broad participation is important, there must nonetheless be a core of working members who take responsibility for organizing and managing the network. It is important that these people represent the larger group. To avoid giving too much power to the few, there must be a continuing process of leadership selection and review by the larger group. This process should be put in place early in the network's evolution.

These leaders must be able to think about the wider needs of the network rather than just what is best for their own institution. They also need enough enthusiasm and time to give to the network. Most importantly, they must stay in touch with a wide variety of the network's members if they are truly to represent their thinking.

Lack of participation

Because networks are voluntary, those who think the network's objectives are very important are likely to spend more time and energy on networking than those who do not. Larger agencies with better educated staff sometimes dominate participants from smaller rural NGOs. Representatives of smaller field-based organizations may be more concerned with their own day-to-day problems and have difficulty in relating to the network's broader objectives (such as environmental concerns). They may feel they have nothing important to add to or learn from many of the discussions. Those speaking local dialects may find it hard to understand or be understood at meetings. These and many other factors may prevent them from actively participating.

Participation is the heart of any network. A successful network sponsors events that encourage the development of shared objectives and provide opportunities for sharing between the entire range of participants—from the small field-oriented programmes to the more vocal national or international organizations. Only activities that are of interest and clear benefit to participants will encourage the enthusiasm and participation needed to make the network successful. The network must produce concrete results if it is to keep the smaller, more practically oriented field organizations interested in collaborating with academic or policy-level participants.

Funding

Informal networks sometimes do not have funds specifically earmarked for network activities. This can limit the activities that can be organized and implemented. If participants must pay for activities themselves, organizations with fewer resources will of necessity be less active than those that can devote resources. On the other hand, when too much money is channelled through a network, competition for funds (or control over them) can lead to favouritism and so to conflicts. Too many resources can also attract freeloaders and fair-weather friends—people looking for travel opportunities, per diems and easy project money, who do not share the network's long-term objectives.

Although effective networks are driven by enthusiasm, in the real world most networks need additional funds for their activities. A structure and process must be set up for planning and administering the network's finances. A donor agency that believes in the network's goals can be approached for support (but should not be allowed to control decision-making). Whether there is a donor or not, operating costs should be kept low and members should contribute at least some of their own resources, for only then will they feel ownership of the network's results. Financial management must be supervised closely by a group of participants who are representative of the network as a whole. Funds must be used wisely if donors (as well as participants) are to continue to trust the network.

Competition

As networks grow, they may come across other institutions working on the same or closely related issues. These other institutions, whether other networks, government agencies or NGOs, may feel threatened by competition from the new network. The competition may not only be for recognition, but also for opportunities, funds and participants.

Competition between groups can sometimes be reduced by forming links between them. Networks working on similar issues can arrange to specialize and to dovetail their activities, enhancing cost-effectiveness. Alternatively, competitive or overlapping networks may be forced by external pressure to merge. Collaboration is usually (though not always) more effective in getting the job done than is competition. Whether or not collaboration is immediately possible, contact with rival groups should be made and maintained so that each can learn what the other is doing, whether functions are being duplicated, and how they might work together in the future.

Time invested in networking

There is a risk that networking may drain staff time and other resources away from field work by becoming an objective of its own. This is especially true of ambitious or outward-looking staff members concerned either to further their own careers or to promote their organization by achieving a high profile in the outside world.

These situations can be avoided by ensuring that networks are managed to serve the needs of members, and that field activities are properly monitored. For these purposes, periodic sessions to evaluate progress and reflect on the network's general direction are essential. These sessions should lead naturally into participatory planning, in which yearly work plans are made and the arrangements for evaluating activities are organized well in advance.

Training for network management

Additional skills may be required for managing networking among professionals. In many cases, formal management training will not be available, however. Many aspects of network management can be learned on the job, or by studying the experiences of others in similar networks. Internships and field visits between networks can be organized for this purpose.

Formal training in network management may be considered an important priority for the future. Initiatives in this area could be taken by networks on the basis of the expressed needs of their constituent sub-networks.

Challenges

Several challenges need to be addressed to enhance LEISA and PTD through networking.

Empowerment of farmers

Active farmer networks provide the essential basis for promoting LEISA. All other types of network should therefore have as their ultimate purpose to promote the establishment and/or support the needs of autonomous farmer networks.

NGOs, researchers and possibly government agencies such as extension services can support farmers' efforts to form networks by documenting and analyzing farmers' experiences with LEISA and by enhancing farmer-to-farmer exchange of these experiences through the funding and organizing of activities such as field days and study tours. Development support organizations can backstop farmers' networks technically by assisting in the planning and design of farmers' experiments and providing information about LEISA practices in other areas. They could also provide training in planning methods and internal organization. Once farmers' networks exist, support services can become more complex, taking the form, for example, of linking different networks, or helping to establish a farmers' newsletter or a national farmers' conference on LEISA.

There is a need for practical guidelines on how to promote the evolution of farmer networks. These guidelines could be based on the documented experiences of NGOs or other development support organizations.

Linking development actors

Although establishing horizontal networks consisting of farmer groups only, NGOs only and researchers only is an important first step, each of these networks needs crucial vertical links with other types of network and service organizations if it is to be fully effective. NGO networks in particular are encouraged to link with international, regional and (sub-)national technical institutions for the provision of support to their field activities. NGOs are in a good position to assess farmers' needs for research and the relevance of existing technologies. Many research institutions now recognize the benefits of relationships with NGOs and NGO networks. A register of networks and service organizations working in the domain of LEISA would be very useful, enabling individual networks to take the initiative in building their own vertical links.

To improve the relevance of field activities, farmer representatives could become members of NGO networks, and NGOs could be represented on research networks. Field-oriented networks could represent farmers' interests by studying the annual reports and other publications of the research centres in their area, assessing the relevance of results for small farmers, periodically visiting the centres and expressing their needs and willingness to collaborate. They could make proposals for collaborative on-farm research and PTD, suggesting research locations and topics. They could also help introduce farmers into the research community.

Learning from networking experiences

The further development of networks devoted to LEISA requires considerable efforts in human resource development. Two initiatives have been identified as essential in this respect, namely the production of a resource book on network management and the development of network management courses within existing training institutes. The book should not be a blueprint on how to network, but should rather contain practical examples and information for use by network managers. It could include a checklist of networking objectives and activities, and options for monitoring and evaluation. There is little documented experience in monitoring and evaluating networks. Developing tools for these exercises requires special attention.

Existing management training institutes could include the management of networks in their curriculum. Regional networks could request tailor-made courses for their member organizations.

Joint activities

Collaboration among different types of network is called for in meeting major challenges in the areas of training and training materials, marketing, advocacy and the assessment of LEISA technologies. Convincing data on the impact of LEISA technologies (on productivity and the environment, and on such socio-economic factors as labour, stability of production, equity, etc) are still relatively scarce. This is a serious problem, because national and international agricultural policies will be modified to enhance LEISA only when its feasibility and advantages are proven. This will require a supra-disciplinary approach, whereas most professionals have been trained as specialists and are operating in organizations with a sectoral mandate. Networks can play an essential role in bringing together the available data on the impact of LEISA technologies.

New methods of research

A change towards LEISA particularly affects researchers and their working methods. While considerable progress has been made in recent years, there needs to be still more interaction between participatory forms of on-farm research and the more traditional forms of station-based research. The process of synthesis between different approaches needs to go several stages further. Research on the management of natural resources is an important new complement to the multidiscplinary commodity focus of efforts at the international level, but its outputs and working methods need to be better defined. Ways of assessing the impact of such research are especially needed. There is room for a vast increase in the number of collaborative projects between research institutions and NGOs, as also between research and extension. And further progress needs to be made in demolishing intersectoral barriers. Networking can be an important tool for pursuing these aims—and particularly for reducing the current tendency of the different groups engaged in research and development to demonize each other. Many research institutions have a long tradition of sharing their experiences with other researchers through networking (see Plucknett et al, p.187; and Smutylo and Koala, p.231). These efforts now need to embrace a wider audience of non-specialists. Likewise, many research institutions have reinforced the user's perspective in their work in recent years (see Prain et al, p.215). However, there is still a long way to go in integrating the research capacities of researchers and farmers. Here too, networks could play an important role in exchanging experiences and providing mutual support.

A challenge for funding agencies

In all cases, networks need funding at some stage of their existence. Seed money may be needed to establish the need for a network and prepare the ground for it. Where network members are not supported by large institutions, money is needed to maintain the network. NGOs need to budget 'networking time' and funding agencies may need to be convinced of the need for what may at first seem an unproductive activity.

References

Arbab, F. and Prager, M. 1991. Searching for alternative systems of agricultural production: The experience of the Rural University. In: Haverkort, B., van der Kamp., J. and Waters-Bayer, A. (eds), *Joining farmers' experiments: Experiences in participatory technology development*. ILEIA/Intermediate Technology Publications, Leusden and London.

Box, L. 1989. Knowledge, networks and cultivators: Cassava in the Dominican Republic. In: Long, N. (ed.), Encounters at the interface: A perspective on social discontinuities in rural development. Wageningen Agricultural University, Wageningen, Netherlands.

Chambers, R. and Jiggins, J. 1986. Agricultural research for resource-poor farmers: A parsimonious paradigm. Discussion Paper 220. IDS, University of Sussex, Brighton, U.K.

Compton, J. L. and Joseffson, K.G. 1993. Farmer networks for sustainable agriculture. Draft, University of Wisconsin, Madison, USA.

Fisher, L. and Moeliono, I. 1992. Good planning can help prevent common problems of networks. In: *World Neighbors in Action 22* (2e).

Haverkort, B., van der Kamp, J. and Waters-Bayer, A. (eds) 1991. *Joining farmers' experiments*. ILEIA/Intermediate Technology Publications, Leusden and London.

ILEIA. 1989. Proceedings of the ILEIA Workshop on Operational Approaches for Participatory Technology Development in Sustainable Agriculture. ILEIA, Leusden.

Nelson, J. and Farrington, J. Forthcoming. Information exchange networking for agricultural development: A review of concepts and practices. CTA, Ede-Wageningen, Netherlands.

OTA. 1988. Enhancing agriculture in Africa: A role for US development assistance. US Government Printing Office, Washington D.C.

Plucknett, D.L., Smith, N.J.H. and Ozgediz, S. 1990. Networking in international agricultural research. In: *Food systems and agrarian change*. Cornell University, Ithaca/London.

Padron, M.C. 1991. Networking and learning. In: *Reflexion* 1 (2).

Reijntjes, C., Waters-Bayer, A. and Haverkort, B. 1992. *Farming for the future*. ILEIA/Macmillan, Leusden and London.

Vincent, F. 1986. Networking strategies: The IRED experience. IRED, Geneva.

Address
Carine Alders, Bertus Haverkort and Laurens van Veldhuizen, ILEIA, Kastanjelaan 5, P.O. Box 64, 3830 AB Leusden, Netherlands.

Strategic Networking: From Community Projects to Global Transformation

by David Korten

Introduction

In March 1988 the Asian NGO Coalition (ANGOC) held a landmark workshop on the theme of strategic networking among non-government organizations (NGOs). The workshop focused on an issue of growing importance for Asian NGOs—the movement beyond village-level projects to a concern with impact at the national level. We identified democratization as a unifying concern for NGO action across Asia. We addressed the need for coalition building at national and sub-national levels to combine resources in pursuit of a shared vision of national development.

The intervening period has been a time of dramatic change for NGOs in Asia as elsewhere. As we broaden our perspectives we realize that the deepening poverty, environmental devastation and violence we see in the villages where we work are not local phenomena. They are pervasive and global—the result of systemic forces that cannot be resolved by action at the village level alone. This awareness helps us to put our work in perspective.

A Self-destructive Vision of Human Progress

Contrary to its promise, economic growth is not alleviating the conditions that underlie the unfolding global crisis. Indeed there is reason to believe that it is the single-minded pursuit of growth that is the cause of the crisis. We face a dilemma. It has become an article of faith among much of the world's population that economic growth is the key to universal prosperity. People the world over expect their leaders to provide it. As the crisis places ever increasing pressures on them, these same leaders, who seldom have time for serious reflection, become increasingly obsessed with the need to take whatever action promises to add to national output statistics in the current year and to fight any action that threatens them. They fail to see that their actions only add to the crisis, not to its resolution.

The favoured short-run policies lead to the concentration of ever greater economic power in the hands of the state and/or large corporate enterprise, each of which is in turn evaluated by society primarily on the basis of its contributions to economic output. In the pursuit of their growth mandate, these institutions seek ever greater control over economic resources, which they mine with an eye only to today's bottom line—usually at the expense of those who are too weak to protect themselves. The greed of the wealthy is indulged while the poor and future generations are deprived of the means of meeting their basic needs and reduced to a struggle for economic survival, stripped of their basic sense of humanity and community. Poverty, environmental destruction and the communal violence that results from a breakdown of the social fabric are all a direct consequence.

Our world is divided not between the developed and the underdeveloped, but rather between the over- and underconsumers of the earth's resources. Because these resources are finite and because total current consumption is at or beyond the ability of the earth's ecosystem to sustain, we are forced to acknowledge that there is a direct link between the behaviour of the overconsumers and the plight of the underconsumers. The despair of the latter cannot be overcome without curbing the greed of the former. The answer lies not in growth, but in a transformation of the values and institutions that define how we use the earth's bounty and distribute its benefits.

Human society is locked into a mind-set that places it on a collision course with the limits of a finite planet and the psychological and social tolerance of its own members. The task before us is one of breaking humanity out of this pattern of collective self-destruction. This task takes us far beyond the traditional role of assisting the poor through village-based development projects. It requires new ways of working and thinking, new organizational relationships, new strategies, and new skills.

How People Change Unresponsive Institutions

The small size and limited financial resources of most NGOs make them unlikely challengers of economic and political systems sustained by the prevailing interests of big government and big business. Yet the environment, peace, human rights, consumer rights and women's movements all provide convincing examples of the power of voluntary action to change society. This seeming paradox can be explained by the fact that the power of voluntary action arises not from the size and resources of individual voluntary organizations, but rather from the ability of the voluntary sector to coalesce the actions of hundreds, thousands or even millions of citizens through vast and constantly evolving networks that commonly lack identifiable structures, embrace many chaotic and conflicting tendencies, and yet act as if in concert to create new political and institutional realities. These networks are able to encircle, infiltrate and even co-opt the resources of opposing bureaucracies. They reach across sectors to intellectuals, the press and community organizations. Once organized they can, through electronic communications, rapidly mobilize significant political forces on a global scale.

Engaging in such processes is a new experience for most development-oriented NGOs. Yet in growing numbers they are joining forces with and learning from the experience of established social movements. As we learn more about the nature of these movements, we realize that they are not defined by organizational structures. They are characterized by spontaneous action that is driven by conviction and given shape and direction by a broadly shared social vision. Participation is ensured through a commitment to certain values rather than by the anticipation of financial or political rewards.

As our understanding grows, we see that strategic networks are the building blocks of social movements. A strategic network is a temporary alliance of individuals and organizations through which their resources are combined in pursuit of shared goals that strengthen the movement's position in relation to major opposing forces. These alliances commonly reach beyond the formal voluntary sector to engage students, media, universities, agencies of government, and responsible business organizations. In many instances they link local, national and international groups.

Many of the participants in a strategic network may be acting on the basis of an immediate agenda or interest, without perceiving themselves to be part of a larger social movement. As is true for the larger movement of which strategic networks form a part, each network may itself consist of countless shifting tactical networks formed around narrower agendas that contribute to the larger strategic objective.

While the success of strategic networks commonly depends on their ability to harness spontaneous voluntary action on a considerable scale, they are seldom in themselves spontaneous creations. Usually one or more individuals or organizations assume critical and highly self-conscious roles as catalysts in their creation, maintenance and direction. NGO experiences in South-East Asia provide rich insights into the nature of this role.

Case of the Nam Choan Dam

In Thailand, the campaign against the Nam Choan Dam, which would have displaced thousands of people and destroyed a major wildlife sanctuary, is an example of a successful strategic network—and one that is especially helpful in understanding the role of the strategic network catalyst. One of the organizations that played such a role in this campaign was the Project for Ecological Recovery (PER), a small environmentally oriented Thai NGO with 10 paid staff members and an annual budget of less than US$35,000. This tiny organization forged an alliance among 38 grass roots organizations in the threatened area, student organizations, conservationists, mass media in Thailand, and an international network of environmental organizations and journalists. The alliance ultimately persuaded the government to cancel the project.

In the course of this campaign PER organized countless meetings and seminars, talked with representatives of government, helped keep all the groups in contact with one another from day to day, organized visits to the dam site by citizen groups from Bangkok and abroad, acted as a clearing house for technical and campaign information, engaged the media, and worked constantly to broaden the campaign's tactics and base of public support.

Each participating group had its own reasons for being engaged. PER recognized and accepted their differing motivations and helped each to find a role within the larger

coalition consistent with its particular commitment and strategic resources. Each group was encouraged to tell its own story from its own perspective, with PER helping each to project the messages to those elements of the public most likely to be sympathetic to its particular appeal:
- The local people faced the destruction of their homes and livelihoods. They were fighting for their immediate survival. PER made them the foundation of the campaign. Rather than attempting to form new community groups, it worked with whatever groups already existed: traditional councils, housewife groups and village scouts. It encouraged different grass roots groups to meet with one another to share concerns, develop joint tactics and form an association, all the while working with existing structures and forces and keeping its own role to a minimum. The local groups staged marches, submitted petitions, applied pressure on their elected representatives and hosted visiting delegations.
- The students, who represented a left-wing perspective, saw the dam as an effort to benefit industrial interests at the expense of the rural poor. They engaged themselves in what they saw as a class war. They arranged demonstrations and publicity events that sometimes involved direct confrontations with government.
- The environmentalists, generally intellectuals and public figures of a more conservative political orientation, were primarily concerned with the preservation of a unique forest. They were encouraged to analyze the environmental dangers and social costs in public fora and at press conferences. Mutual suspicion of one another's motives between this group and the students led PER to work with each group separately on its own terms rather than trying to bring them together to plan joint action.
- The journalists and film makers were engaged through seminars, press releases, site visits and letter writing campaigns as professionals whose role is to keep the public informed on important issues. They disseminated information about the uniqueness of the area to audiences in Thailand and abroad and publicized the consequences of the project for people living in the area.

The PER's mode of working is quite different from that of the typical development-oriented NGO. For example:
- It maintains a low profile, never using its own name, always projecting the image of other groups and highlighting their commitment to the cause.
- It does not take on any function that another group can perform, confining itself to facilitating links and filling temporary gaps not serviced by other organizations.
- It does not fund local groups, preferring to strengthen their self-reliance by helping them plan and carry out their own fund-raising events, such as rock concerts.
- It does not publish a newsletter. When communicating with a particular constituency it uses that group's own newsletter. When it wants to get a message out to the public, it uses the mass media. Thus it reaches a far larger audience at a fraction of the costs typically incurred by the average institutional information department.
- It keeps itself small and its budget modest, working by activating, enabling, amplifying and focusing existing social forces. It lives by the logic that big organizations have to take

on large conventional projects to justify themselves. They necessarily become competitors with other organizations and interests rather than facilitators who measure their own success in terms of their effectiveness in helping others to be strong and successful.

PER's experience suggests a number of useful operating principles for organizations serving as a strategic network catalyst (Box 1).

Box 1. Operating principles for strategic network catalysts

- Maintain a low public profile. Emphasize the commitment and contribution of other organizations to the network's goals. Measure own success by effectiveness in making others stronger and more successful contributors to these goals.
- Recognize the differing motivations and resources of the groups engaged in the network.
- Look to those who have the most direct and compelling interest in the outcome to provide leadership.
- Continuously scan the environment for opportunities to engage new participants who bring new perspectives and may appeal to additional segments of the public.
- Do not take on any function that another group can perform. Facilitate links and fill gaps not serviced by other organizations.
- Work through existing communication networks and media to reach large audiences efficiently.
- Help other groups find their own sources of funds, but don't become a funder.
- Keep own staff and budget small to ensure flexibility, avoid competing institutional interests and maintain dependence on the effective action of others.
- Use protest actions to position the movement to advance a pro-active agenda.

PER's experience with a variety of campaigns against socially and environmentally harmful development projects has led it to the conclusion that attacking ill-conceived projects is not enough. It now realizes that the government's need for dams stems from policies that fail to account adequately for their residual social and environmental costs. It must put forward credible alternatives. For example, the dam projects that PER has successfully fought are being put forward by a government desperately seeking ways to meet the projected demand for energy resulting from Thailand's export-led growth strategy. There is a need not only for an alternative energy policy that emphasizes energy conservation and efficiency, but also for an alternative development policy that points to new ways of enhancing the well-being of the Thai people, without significant dependence on large new centralized power generation facilities.

Producing and presenting credible alternatives poses a serious new challenge to PER. Government, business and existing technocratic think-tanks are all committed to established concepts of development. PER thus sees a need to develop its own research capacity as it redefines its role from one of resisting harmful development projects on a case-by-case basis to one of formulating and popularizing well thought out alternatives more beneficial to the long-term interests of Thailand's people.

Theories of Poverty: From Basic Needs to Development Vision

Tim Brodhead says that to be a development organization it is essential to have a theory of poverty that directs us to its underlying causes. Without such a theory the organization inevitably remains a relief and welfare agency, responding only to poverty's most evident symptoms.

Many NGOs concerned with the plight of the poor did indeed begin as relief and welfare organizations, and many remain so today. They see that people are unable to meet their basic needs and, without asking why, respond in the most direct and immediate way— by providing food, clothing, health care and shelter as required. They engage in first-generation strategies.

The more thoughtful NGOs at some point find themselves asking, 'Why are these people poor?'. They begin, at least implicitly, to formulate a theory of poverty. They attempt to 'look upstream', searching for the source or cause of the problem. Many NGOs that pursue this question conclude that the problem is local inertia, a sort of self-imposed—and by implication self-correctable—powerlessness resulting from lack of organization, political consciousness, belief in self, credit and basic skills. Armed with the belief that this inertia can be broken through appropriate external interventions, they set about to intervene with community development programmes. They reorient themselves to second-generation strategies.

When the theory of community inertia proved to be inadequate, some of us looked further upstream. This led to the realization that in large measure the powerlessness of the village community is not self-imposed. Rather, it is externally imposed and sustained by policies and programmes, often originating from the state and funded by foreign agencies, that deprive the poor of access to productive resources and maintain them in a state of dependency. Development projects, such as dams and industrial forest plantations, that displace the poor from their homes and means of livelihood are among the most obvious examples. In response, some NGOs have set about lobbying for changes in government policies and working with government to reorient its programmes in ways that strengthen rather than undermine local communities.

Some NGOs are now taking another look still further upstream. What they see is deeply disturbing. Many of the most devastating programmes and policies are a direct consequence of the way human society has come to define development itself. They are embedded in a growth-centered development vision and are central to the institutions that we have collectively created to pursue it. We are now looking at the most fundamental driving forces of the global system and coming to realize the extent to which the poverty, environmental destruction and communal violence experienced in the villages of Asia are symptoms of forces that have locked human society onto a self-destructive path that ultimately threatens the very survival of human civilization.

Many NGOs have become expert in consciousness-raising at the village level. They had defined the problem as one of an inappropriate mind-set. Now we see that though the problem was correctly defined, its scope was seriously underestimated. Consciousness-raising is essential, but not only for the poor villager. It must be universal, including the power holders in global society.

As we reflect on the events of the past few years, we will see that we have been engaged in a continuous process of what Hazel Henderson calls 'upstreaming', reaching beyond the evident consequences of the problem at hand to address its source. We are finding that this is not a simple matter. Each time we move upstream we find the issues are more complex and the vested interests more powerful. We feel less confident in our traditional skills and face the need to create ever wider networks. We are led into increasing involvement with the global context of our national political and economic systems and pulled into ever larger and more complex international coalitions.

To achieve changes of the scope and magnitude required, it is necessary to think of the NGOs' people-centered development alternative not as a village project but as a global people's movement for social transformation. The strategic networks we know of are among the countless such initiatives that are giving this movement its vitality and direction. They represent, however, only a bare beginning. On a global scale thousands more are needed, each with their own catalysts.

Issues

Many issues need to be addressed by those of us who define our role as a catalyst in the formation and guidance of strategic networks as elements of a larger movement for global social transformation. Here are some of them:

From protest to pro-action

Engaging major constituencies in protest can make an important contribution to strengthening awareness of issues and building commitment to activism. Protest is relatively easy to organize as it is usually easier to build consensus about what should *not* be done than about what constitutes a positive alternative. At the same time protest actions only pose barriers to the negative forces of the growth-centered development vision. Even when successful, they do not transform those forces.

Eventually there must be attention to building support for a pro-active agenda aimed at transformation. The question should be continuously in mind: What do we want in place of what we cannot accept? Each protest action should be consciously designed to lead toward defining such alternatives and building supporting constituencies.

Building citizen democracy

Democratization is a key theme of NGO activity throughout Asia. We have learned, however, that the institutional forms of democratic governance are in themselves little more than empty shells. They provide citizens with the means to engage in the governance process, but there is no guarantee that citizens will use them. Unused, they will inevitably be abused by those in power. Our experience is leading us to the realization that 'democracy is not something we have; it is something we do'. Democracy cannot exist without citizen action.

Democratization is best thought of as a process of building capacities for and commitment to citizen action, through action itself. The traditional organization building agenda of NGOs is a part of this process, but only a part. Strategic networks are another. Single organizations are rarely successful in taking on significant policy and institutional change agendas when they act entirely on their own. Where the issues involve significant external political and economic forces, strategic networking becomes an essential mode of action.

Strategic networks are important training grounds for, as well as instruments of, citizen democracy and should be treated as such. Strategic networking can also contribute to the essential process of rebuilding a sense of community, recreating the social structures based on shared values that modern society, with its emphasis on impersonal market transactions and hierarchical organizations, has disrupted. It is a means of helping to break the alienation and powerlessness that this disruption has left behind.

In a vital democratic system, an individual person or organization may be engaged simultaneously in several strategic networks that involve different agendas and different combinations of actors. Citizen democracy grows out of the thickening web of active networks that form around a growing number of needs and issues.

Each strategic network effort should seek consciously to leave behind a more informed and active citizenry with a strengthened sense of belonging to a local, national and global community of caring citizens. Though networking links are often ephemeral, each time a link dissolves there should be left behind a positive memory trace that will make new links easier to form.

Forming alliances across social movements

The NGOs concerned with poverty or social justice present an exceedingly weak force in the face of the transformation agenda. Remaining isolated and even competitive with one another and focusing almost exclusively on micro-level issues, it is stretching the meaning of the term to call them a movement. Fortunately, many of them are moving quickly beyond their traditional limitations and are learning the ways of strategic networking, often through alliances with organizations from movements more experienced in that mode of action.

We are coming to realize that the people-centered development or social transformation movement is in fact a meta-movement that embraces the pro-active agendas of many existing social movements, including the environment, human rights, peace, women's, social justice and consumer protection movements. The meta-movement will emerge as a truly significant force for change only as the participants in its component movements come to recognize the extent to which the realization of their own agendas depends ultimately on achievement of the larger transformational agenda. There are already plenty of examples of actions that join the interests of two or more of the component movements. Links between the environment, human rights and social justice movements seem to be particularly common. Encouraging and strengthening these tendencies is highly desirable.

Activism versus service provision

There has been a tendency to group together all NGOs concerned with the needs of the poor into a single category. They join the same coalition bodies, have the same kinds of registration, use the same terms to identify themselves, attend the same training programmes, often seek funds from the same donors. But are they the same?

While still too early to say for sure, there are indications that we may be seeing a parting of the ways—a permanent division among NGOs between those that choose to provide specialized services (elsewhere described as public service contractors) and those that choose to be social activists, working in more catalytic roles. These two roles require fundamentally different skills and orientations.

Service providers have a natural, and probably appropriate, tendency to grow in size and service area and are more likely to have hierarchical organizational forms. Their activities feature routine modes of working that are unlikely to be offensive to the donors on which their survival depends.

The activist organizations are likely to have a smaller staff, although they may have large membership bodies organized around decentralized local chapters. They will also tend toward looser, more decentralized structures. Very little of what they do could be considered routine. They are constantly facing new challenges. They do not necessarily seek political controversy, but are at the very least forced to accept it as an occupational hazard. Their funding is likely to be fairly precarious and they are more likely to depend on voluntary contributions of funds and time.

These thumbnail sketches are based more on a theoretical analysis of differences than on an examination of actual experience. In such an examination we might look for answers to more specific questions, such as the following: What are the characteristics of effective strategic network catalysts? To what extent have organizations with a service delivery programme been effective in a catalytic role? Can catalytic organizations be effective participants in service-oriented networks? To what extent do the activist organizations feel compelled to look and act like service providers in order to attract funding and maintain their acceptability to government? Is it productive to blur the distinctions between the activists and the service providers? Or does the lack of a clear distinction hinder our ability both to develop effective national service delivery systems and effective fora for the practice of citizen democracy?

NGOs no longer enjoy the luxury of being inconsequential actors at the periphery of the development stage. Our choices make a serious difference to the global future. We must take our responsibilities seriously and prepare ourselves accordingly. We are only beginning to understand the nature of strategic networks and the critical roles of the catalysts that give them shape and direction. We must make rapid advances in that understanding, in the development of our skills as effective catalysts, and in sharing our insights with others who might assume similar roles.

Source
Reprinted from: *Forests, Trees and People* Newsletter 18 (September 1992), Forests, Trees and People Programme (FTPP), Uppsala, Sweden.

Address
David Korten, People-Centred Development Forum International Secretariat, 14 East 17th Street, Suite 5, New York, NY 10003, USA.

PART TWO

Farmers' Networks

Farmers' Networks: Key to Sustainable Agriculture

Bertus Haverkort

Introduction

Not only is there considerable variety in the origin, goals and organizational structure of farmers' networks, but these networks are also continuously adapting themselves to changing circumstances. Recent changes in the economic, agro-ecological and political environment have led to a new flexibility and dynamism in farmers' attitudes. This is evident both in their practices at village level and, through their organizations, in their responses to challenges at the national, regional and international levels. Conventional development approaches have tended to use farmers' organizations and networks to increase their control over the development process, influencing them in the direction deemed desirable by outside agencies.

This paper advocates a development support policy whereby farmers' networks are seen and respected as a way in which farmers can take control of the agricultural development process. Outside support, if required at all, would be given only to further empower them.

New Interest in Farmers' Networks

Farmers are the 'carriers' of agricultural development. They have their own approaches to the development and application of technologies, and their own patterns of decision-making. While farming itself is done mainly on individual farms, the broader rural community plays an essential role in farmers' strategies for survival and development. Rural people, like city dwellers, like to get together to share information and other forms of mutual support with others whom they trust. There is nothing new under the sun, and networking between farmers is as old as farming itself. Yet a renewed interest in farmers' networks can be detected.

Compton and Joseffson (1993) state that this new interest can be seen as a reaction to two major social problems of modernizing societies. First, increasing fragmentation of society as a result of commercialization, specialization and large-scale production leads to the loss of a sense of community. This leads to the adaptation or revival of traditional forms of information exchange and cooperation, or to the emergence of new ones. Second, people start to rebel against human service systems such as bureaucracies and businesses, especially if they feel that these have largely failed them or work against their interests.

New forms of farmer networks are thus arising out of a need to gain access to power, new technology and information relevant to the farming community. Established research and extension systems have, until recently at least, had little relevant information and advice to offer small-scale family farmers. These farming systems did not fit the way most researchers and extensionists perceived agricultural development and progress (specialization/monoculture, chemicalization, mechanization and large-scale production). The large-scale farmers who fit the establishment's view of agricultural development were seen as progressive. It has repeatedly been demonstrated, both in the North and in the South, that agricultural policy instruments such as subsidies, price systems, legislation, research and extension have been designed and used predominantly to benefit these progressive farmers. By giving priority to these farmers, policies have neglected the needs and damaged the potential of smaller farms. The rationale for this selective policy is reflected in the frequently heard admonition, 'get big or get out'.

In spite of being deprived of support, small-scale family farms have persisted, and the farming strategies adopted by their owners and operators remain viable. The long-standing deprivation faced by such farmers is a major driving force behind their new determination to draw upon each other's strengths. Farmer networks enable them to share their experience. The cooperation they encourage offers them the opportunity to create alternative development models and to influence policy.

Potential Benefits of Farmer Networks

It is important to recognize who benefits from farmer networks. In addition to the participating farmers, the rural community as a whole benefits from the improvements in agriculture promoted by a network's influence.

Agricultural research and extension organizations can benefit in two ways. If they establish a relationship of confidence with the network's farmers, they will gain a local partner that will participate creatively and critically, bringing a qualitative improvement in the research programme. In addition, farmer networks can take on more responsibility for local agricultural experiments and demonstrations, helping to disseminate the results of research and so bringing a quantitative gain in impact. In both cases, the cost-effectiveness of research is improved.

Compton and Joseffson (1993) mention the following advantages of farmer networks:

Risk sharing
A basic function of farmer networks is to build confidence among member farmers and to provide support and encouragement in risk-taking. Especially in risk-prone areas, it is common for families to look after each other in difficult times. Self-initiated and self-directed farmer networks can provide a safety net and a buffer at such times.

Sharing experiences
New farmers can learn from older farmers and inexperienced farmers can learn from experienced ones. All farmers can learn from each other and so avoid the unnecessary repetition of mistakes.

Experimentation and demonstration
The experiments conducted by farmer networks can effectively and efficiently serve to develop farming practices that respond to local conditions. The network can assign different parts of an experiment to different members. This avoids duplication and enables farmers to investigate a proposed new practice more completely and more quickly. Cooperation in the network will help to improve the design of farmers' experiments and to develop farmers' research skills. When experiments and demonstrations are carried out by farmer networks they naturally include appropriate consideration of risk, labour requirement and community values—factors which researchers working by themselves have often had difficulty in building into their programmes.

Networks allow participating farmers to discuss and analyze each other's observations and experiences. This process results in valuable research questions. When forwarded to agricultural research organizations, these questions and requests should, presumably, carry more weight, because they are put forward by a network rather than an individual farmer.

Extension and communication
In addition to generating and exchanging knowledge based on farmers' experiences, farmer networks can obtain and disseminate agricultural information from outside the network. They can serve as a link not only between individual farmers but also between farming communities and the agricultural extension system. Networks have often emerged in response to the absence of an adequate extension service. Yet the existence of such networks can facilitate the work of extensionists and researchers provided these accept the network for what it is—namely a forum for the articulation of collective needs.

Empowerment
Farmer networks can focus around many areas of common interest and needs. As farmers join together and begin to support and learn from each other, a network develops strength. It becomes increasingly able to command respect and attention, and to promote the

common interests of its members and the larger community. Practical outcomes can be cooperative purchasing of supplies, cooperative selling, and marketing of produce. Well established networks can become effective advocates of policy change, claim improved access to public services for their members, and help to enlist public sympathy for, or at least interest in, the issues of environment and development which affect farmers' lives.

During a workshop on networking for low-external-input agriculture held in the Philippines in 1992, it was generally agreed that the most essential ingredient for the promotion of low-external-input and sustainable agriculture is the existence of strong farmer-based networks in the rural community. Development support networks, such as those of NGOs or of research institutions, should therefore aim to cooperate with and/or support the needs of farmer-based networks. The first step in this direction is to understand how farmers' networks operate. Recent research has provided important insights in this area, which is one that is frequently overlooked by development organizations.

Indigenous Knowledge and Communication

Warren and Cashman (1988) define indigenous knowledge as the sum of experiences and other forms of knowledge of a given ethnic group that forms the basis for their decision-making. Farmers have sophisticated ways of looking at the world. They have names for many different kinds of plants, ways of diagnosing and treating human and animal diseases, and methods for cropping both fertile and infertile soils. This knowledge has accrued over many centuries and is a critical and substantial aspect of the culture and technology of any rural society. Indigenous knowledge about agriculture is intimately connected with knowledge in other spheres of life, notably health, social systems and spirituality. It is being preserved, communicated and changed. The last point is important: indigenous knowledge is not static, but is continuously adapted to meet the changing needs and circumstances of the rural population. In this process, networks play important roles.

Thrupp (1989) stresses the importance and unique value of indigenous knowledge, but warns against romanticizing its potential. The type, extent and distribution of knowledge varies greatly in different societies, but all resource-poor farmers have valuable knowledge. The capacities of individuals to innovate (create new knowledge) and to apply and transfer existing knowledge are also diverse. She concludes that indigenous knowledge continues to be marginalized by the dominance of top-down development approaches, the pressures exerted by agrochemical firms, scientific professionalism and other political and economic forces. She advocates enabling people to establish legitimacy of their knowledge for themselves as a form of empowerment.

McCorkle et al (1988) showed that, in Niger, knowledge exchange takes place in more or less regular ways in a wide range of places. Informal networks frequently emerge spontaneously around traditional institutions such as markets, village wells, grain mills, blacksmiths' workshops, health centres and churches/mosques. Funerals, festivals, rituals and tribal meetings may all be occasions for the exchange of knowledge. Within these traditional institutions, predetermined roles are given to certain persons. At markets, specific areas are frequently reserved for certain commodities, where producers, traders and customers meet to exchange goods, money and information. A market can thus be seen as a conglomerate of different networks, each focusing on a specific aspect of the indigenous knowledge system. Village wells are known to be places where women exchange information. Religious meetings and rituals are often led by spiritual leaders, and offer the opportunity to exchange experiences and develop community consensus on important local issues.

Influence of the cultural environment

Schuthof (1990) has shown that, in Zimbabwe, within a specific farming community there are at least two different types of network: those of Christians and those of non-Christians. Whether a person is a Christian or a non-Christian to a large extent determines the way he or she views the 'management' of nature. Within traditional society the ancestors and spirits play an important role in determining the success or failure of agricultural production. The rainmaker is a kind of medium between the spirits and the farmers, and is regularly consulted on agriculture-related activities and decisions. The introduction of Christianity in the 1960s resulted in considerable change. Christians do not offer beer to the ancestors. Nor do they consult the rainmaker for blessing the seeds. Participation in traditional or modern social activities was also an indicator of whether one asked the government extension agent or the rainmaker for advice on agricultural problems. Schuthof found that there was hardly any information exchange between the rainmaker and the extension worker. Both claimed to know how plant diseases should be eradicated, but the logic of their respective knowledge systems differed radically. Schuthof concluded that rural peoples' knowledge networks were embedded in a cognitive framework that was religous as well as social and economic in nature, with the result that farmers' rationales and goals were essentially subjective and varied substantially among farmers. This cognitive framework determined the way in which farmers give significance to their lives. Belief in the Christian God or in the spiritual world of their ancestors determined the network to which farmers belonged.

Similar observations were made by Haverkort and Millar (1992) in their research in Ghana. They found that in traditional society the priests, soothsayers, elders, village headmen and chiefs played important roles in local experimentation. New ideas could be tried out and changes made as long as the Gods' consent was sought by the local priests, through the ancestors and after consulting the soothsayers. Sacrifices were made and traditional rituals performed before experiments were conducted. Here, the traditional institutions had an important regulating function. The outcomes of experiments were indicated not only in terms of yield or economic returns, but also by other signs of the Gods or ancestors, such as health or accidents. In Ghana, the traditional institutions that connected the different people responsible for local governance and decision-making and allow for the rituals to take place formed an important network. These institutions also played a role in the exchange of information and in farmers' experiments. The authors concluded that there was a great lack of communication between extension agents and researchers on the one hand and indigenous leaders on the other. They advocated an approach whereby outside agencies established contacts with indigenous institutions, so that the knowledge, resources and influence of both parties could be combined in a mutually beneficial way.

In his paper on indigenous knowledge in Bolivia, Rist (see p.93) describes similar mechanisms for farmers in the Andes. The indigenous cosmology invokes both the natural and the spiritual worlds, which are intimately related. Both worlds are worthy of attention. Rist also describes how this traditional knowledge system is being eroded by the influence of outside agencies such as the school, religion, non-government organizations and rural extension, and by factors such as temporary migration, emerging new needs and food donations. Among the many connections re-established by learning once again to set a value on this knowledge are those between different agro-ecological zones, as illustrated in the in-situ conservation and exchange of germplasm and in the exchange of knowledge about erosion control.

Pereira and Seabrook (1990) describe the indigenous knowledge system of the Warli tribe in India, illustrating its richness and appropriateness through many examples. Here too, knowledge is embedded in, modified by and transferred through indigenous networks. The Warli's agricultural experiments are part and parcel of community activities. The authors go further than most anthropologists in concluding that outside development agencies need to have faith in the validity of the principles on which the traditional system is based. This requires a radical change in agencies' ways of thinking and values. Indigenous knowledge is often not seen as valuable and valid in itself, but merely as something to be taken into account when introducing Western concepts of development. Enforcing conformity with mainstream beliefs through education, health, agricultural and other services based on modern science and technology is a sure way of destroying indigenous knowledge systems. According to Pereira, modernization, if necessary at all, should be an adaptation of good traditional principles and methods to the problems of today, with inputs from new science and technology used only if they are not seen as destructive.

Power, politics and progress
Bebbington (1991), van der Ploeg (1990) and Toledo (1992) draw attention to the power and political aspects of indigenous knowledge and knowledge networks. Bebbington

describes farmers' organizations in Ecuador, where indigenous rural peoples are formally organized at a variety of levels. Community-based organizations federate into second-order organizations at parish, county or provincial level. Often, these organizations were originally formed to campaign for land rights. Later they became active in implementing rural development programmes. The struggle for land has often been presented as a struggle for the right to protect and recover a traditional way of life. For many farmers' organizations in Ecuador, the steady modernization of indigenous societies is not merely a matter of technical change but part of a process of cultural assimilation into a dominant society, that is to be resisted. Bebbington describes a meeting of indigenous peasants' federations that made a collective declaration denouncing the activities of the 'so-called' extension agents of the state as 'cultural aggression' and 'instruments of manipulation', used by the dominant Hispanic society to subjugate them. Some organizations argue that indigenous social and technological practices should be researched and reinstated. These practices demonstrate the feasibility of an indigenous way of living, producing and organizing. Indigenous knowledge is also a symbol of and tool for resistance to socio-cultural assimilation. Chiriboga (quoted in Bebbington) argues that the concern to identify an alternative model of development based on local resources, local forms of social organization and indigenous practices is inspired by a conscious effort on the part of the peasants' organization to distance itself from the pressures resulting from dependence on the market and from the policies associated with structural adjustment.

Bebbington also draws attention to another form of reaction to modernization: selective modernization or resistant adaptation. This occurs when traditional technologies cannot generate sufficient local income to prevent out-migration. The introduction of cash crops and yield-increasing technologies is seen as a strategy for accumulating resources that will reduce migration and help to maintain other traditions such as dress, language and organizational forms.

Toledo (1992) observes similar changes in Mexico. Farmers' movements that focused initially on land rights and farmers' control of the production process now fight for the defense of nature, for the survival of the very ecological system on which their livelihood depends. This new focus leads to a revaluation of traditional concepts of the relationship between man and nature, in which notions of respect for nature, reciprocal maintenance, self-sufficiency and equality prevail. These changes are likely to mobilize national and international support for these movements.

A great number of politically oriented farmers' organizations are reported to exist in Latin America. And there are powerful examples from other continents too. The Chipko movement in the Himalayas was started by village women in the Reni forests of the Chamoli District of Uttar Pradesh, India, who wished to prevent trees from being felled. The movement spread to many remote areas of India, Nepal and Bhutan, and has successfully brought commercial forestry to a standstill, stopped the construction of several large dams and greatly reduced unregulated mining. Guha (1989) emphasizes that Chipko is the successor of earlier peasant struggles in the area, but at the same time it goes beyond them. It has made a dynamic contribution to the public debate on the environment, both in India and abroad.

Linking with Indigenous Knowledge: Opportunity or Risk?

Van der Ploeg (1990) sees as a dominant and central feature of modern agricultural science that it systematically expropriates farmers' knowledge and therefore their control over their working and living conditions. Science and technology rupture existing development patterns and indigenous systems of learning, experimenting and teaching. The knowledge of farmers is made superfluous and their labour is subjected to external interests and perspectives. It is this that, increasingly, makes farmers say: 'Up to here and no further'. They draw up a line of defence to protect their own essential interests and perspectives. Their local knowledge is the primary weapon they use in self-defence. Farmers' knowledge is not just a neutral set of items to be exchanged for or blended with other forms of knowledge. It is the key to their sense of identity and power, and therefore to their survival.

Van der Ploeg questions whether, through simple expedients such as networking, on-farm research and participation, the profound contradiction between farmers' knowledge and scientific knowledge can be superseded. Should not the struggle to consolidate, reinforce or even reinstate local knowledge be oriented primarily towards increasing farmers' power and independence, thereby creating better conditions for the further development of indigenous knowledge?

I feel that this question addresses the heart of the matter. The current renewal of interest in indigenous knowledge is both an opportunity and a risk. As shown by O'Brien and Flora (1992), focusing on indigenous knowledge can further empower rural communities, but it can also—and this despite the good intentions of development workers—lead to a further sell-out, preparing the way for further control of rural communities by outsiders. Despite the rhetoric of empowerment, development agencies (multilateral, bilateral and non-government) have generally failed farmers in their support. Much farming systems research continues to aim at increased use of purchased inputs and Green Revolution technologies.

In short, I join van der Ploeg, O'Brien and Flora in raising the question: Will support to farmers' networks lead to the further appropriation of indigenous knowledge and further control by outside agencies, or will it lead to greater empowerment of farmers?

Documentation of farmers' knowledge
The international development set has recently given much more attention to indigenous knowledge. At grass roots level several experiences have now been recorded in which

development initiatives began with an inventory of existing farmers' knowledge and subsequent activities were built on this. Cases of documenting farmers' knowledge for the purpose of empowering farmers have also been described.

Pereira and Seabrook (1990) give an example of a 12-year-old girl of the Warli tribe in India who knows the names of over 100 herbs, shrubs and trees and their varied uses. She knows which plants are a source of fibre, which are good for fuel and lighting, which have medicinal use and which can supplement the basic diet of cereals and pulses with essential proteins, vitamins and minerals. She possesses a vast, complete knowledge system on animal husbandry, agriculture, meteorology, herbal medicine, botany, zoology, house construction, ecology, geology, economics, religion and psychology. Of course, not all of this knowledge is valid in Western eyes: the authors assert that some superstition is undoubtedly involved, and a few practices are positively harmful. But this is a tiny fraction of the Warli's 'science', and in general they have the wisdom to use their knowledge well.

McCorkle et al (1988) describe some 20 case studies of successful farmers' innovations. These include the introduction of short-cycle millet varieties, new land preparation methods, the construction of mini-catchments, seed pocket manuring, dry-season gardening, forage utilization, biological pest control, and a range of ethnoveterinary medicines. The authors found that farmers in Niger are open to seeking out and applying new agricultural ideas. They can plan, implement and evaluate on-farm research trials, and demonstrate a sophisticated understanding of the complex interactions among the many variables they manage. The case studies also show that there is a rich body of local technical knowledge in Niger's agriculture that could be useful to farmers throughout the Sahel. Farmers choose technologies because these reduce risk, generate income, are affordable and readily available, save labour and fit into current farming practices. Research and extension offer very few technologies deemed appropriate by farmers. The authors recommend efforts to strengthen farmer-to-farmer communication of indigenous agricultural knowledge, offering farmers more opportunities and incentives to experiment for themselves and strengthening farmer feedback loops in research and extension.

The case of the Agroecology Programme of the University of Cochabamba (AGRUCO), described by Rist (see p.93), shows how the documentation of indigenous knowledge can yield important new insights into feasible technical options. Field workers collaborate with farmers to develop handbills on topics such as soil management, mixed cropping and weather forecasting. These handbills serve to prevent the disappearance of knowledge on these topics and to redisseminate useful technology to the community. Such efforts can also be the starting point for joint research. In Peru the Andean Project for Indigenous Technology (PRATEC) has documented over 600 Andean technologies and has made these available for farmer-to-farmer communication.

Gata and Kativhu (1991) provide an overview of indigenous farmers' knowledge in Zimbabwe. They describe indigenous ways of conserving and managing natural resources (trees, wildlife, soil and water) and crop production. They conclude that women in particular have developed a sound scientific and technological knowledge system for agriculture which should be integrated into formal knowledge systems.

However, many examples show that documenting farmers' knowledge may have a questionable effect. The growing international emphasis on biotechnology and biodiversity has led to increased research in ethnobiology—the study of indigenous peoples' knowledge of local plants, animals and biological processes. This research is likely to yield more academic degrees for the students who conduct it than useful technologies for the resource-poor farmers owning the raw materials under research. Worse still, samples of such farmers' materials can easily be used by commercial organizations to identify and isolate active ingredients, mass reproduce them and commercialize and/or chemicalize production. In some cases commercial organizations even patent the products they have developed on the basis of such research. An example from Christie (1993) shows how farmers' intellectual property rights need protection:

> The African soapberry, or *endod*, has been cultivated by Ethiopians (and other Africans) for generations, for use as a laundry soap and shampoo, and to stun fish. While doing field research, an Ethiopian scientist noticed that the soapberry also seemed to kill freshwater snails downstream from laundry and bathing sites. He spent 29 years collecting and conducting research on more than 400 varieties of the soapberry plant, and on schistosomiasis. He identified a few soapberry varieties which are particularly effective against the snail that carries schistosomiasis. American scientists wondered if *endod* would also kill the zebra mussel. It is an import into the Great Lakes water system of Canada and the USA which causes multi-million dollar damage by blocking water intake pipes in the Great Lakes waterway. *Endod* did kill zebra mussels, and the University of Toledo has applied for a patent for any applications of *endod* used in controlling the zebra mussel. The Ethiopian scientist would receive a portion of any royalties paid when a commercial product is developed. But what of the people who originally cultivated and used *endod*? Without them and their knowledge, none of this subsequent research would have happened.

Farmer networks may help to protect farmers' intellectual property rights somewhat better in future. To avoid expropriation, indigenous networks will need to guard their knowledge carefully. Yet there is an intrinsic conflict of values between many rural societies in developing countries and Western industrialized society: land, water, genetic resources and knowledge are mostly seen by the former as common property, for which systems to protect property rights would be inappropriate. This makes them highly vulnerable to expropriation, as the colonial and neocolonial eras have made painfully clear. An essential element of future outside support to indigenous knowledge systems may therefore be help in establishing or supporting local systems for the protection of property rights.

Although the same indigenous knowledge is often widely shared, different individuals or groups may have different degrees of access to specific subject areas. Knowledge is then a source of power or privilege. Juma (1989) found that in Eastern Kenya only certain elders of certain tribes know certain aspects of medicine and the collection sites of valued plants. These elders have high status because of their secret knowledge.

Combining indigenous with outside knowledge can be very fruitful since both, in isolation, have their relative strengths and weaknesses. But the most important condition for success is the willingness of outsiders to recognize the authority and cultural values of farmers. This recognition does not mean making an inventory of useful knowledge in the hope of capturing something of economic use in the outsiders' world. It means turning over the right to direct the development process to farmers and entering into a dialogue with them of which the outcome is not certain.

Bridging the gap between related networks

Farmer networks do not exist in isolation. Box (1989) analyzed the knowledge network on cassava in the Dominican Republic. He concluded that there were four or five different networks for this crop alone: farmer networks, research networks, extension networks and networks of traders and coordinators of development projects. Although these networks co-existed in the same area and referred to the same crop, there was very little interaction between them. Among these different networks the values and priorities of members differed considerably and the information exchanged referred to different aspects of the crop. There was even a tendency to disqualify other networks in priority setting and problem formulation. This led to the waste of opportunities for cooperation and crop development. Box describes his experiences in bringing together the different networks in workshops. This presented farmers, researchers, traders and extensionists the opportunity to learn from each other and to start or intensify cooperation. Such an approach may be highly cost-effective and illustrates a useful role of outside development agencies in working with farmer networks.

Linking farmers' networks

The success of the Campesino a Campesino Movement in Central America (see Holt-Giménez and Cruz Mora, p.51) is partly due to farmer exchange between countries. Farmer leaders of one country visit another and receive encouragement and technical support to launch farmers' experiments in their own country and community. Among the important mechanisms for strengthening the voice of the South is the organization of South-South exchanges of experience and knowledge.

The Western approach to agriculture currently faces many problems related to equity and the environment. In these areas Northern organizations could learn a great deal from the South. The same applies to women's networks and to the ethical and intellectual values they represent—much can be learnt from these by men.

In efforts to redesign the agenda of agricultural development, the contributions of these underutilized resources may be essential. Networking at all levels is therefore of the utmost importance.

Conclusion

Successful networking, whether in purely indigenous forms or in linking different types of network, depends to a large extent on people 'daring to share', instead of using the 'net' to catch a prey.

References

Bebbington, A. 1991. Indigenous agricultural knowledge systems: Human reflections on farmer organizations in Ecuador. In: *Agriculture and Human Values* 8 (1 and 2). Gainesville, USA.

Box, L. 1989. Knowledge, networks and cultivators: Cassava in the Dominican Republic. Research report. Wageningen Agricultural University, Wageningen, Netherlands.

Christie, J. 1993. Behind the promise of biotechnology. Paper presented at the Conference on Prioritizing the Biotechnology Agenda for Zimbabwe, March 1993, Harare.

Compton, J. L. and Josefsson, K.G. 1993. Farmer networks for sustainable agriculture. Draft. University of Wisconsin, Madison, USA.

Gata, N.R. and Kativhu, A.L. 1991. Research and documentation of indigenous science and technology for sustainable agriculture: Good systems and natural resource management in Zimbabwe. DR and SS, Harare.

Guha, R. 1989. *The unquiet woods: Ecological change and peasant resistance in the Himalaya.* Oxford University Press, Delhi.

Haverkort, B. and Millar, D. 1992. Cosmology, peoples' science and indigenous experimentation. Paper presented at the International Conference on Ethnobiology, November 1992, Mexico. ILEIA, Leusden.

Juma, C. 1989. *The gene hunters.* Zed Books Ltd, London.

McCorkle, C.M., Brandstetter, R.H. and McClure, G.D. 1988. A case study on farmer innovations and communications in Niger. Academy for Educational Development Inc., Washington, D.C.

O'Brien, W.E. and Flora, C.B. 1992. Selling appropriate development vs selling out rural communities: Empowerment and control in indigenous knowledge discourse. In: *Agriculture and Human Values* 9 (2). Gainesville, USA.

Pereira, W. and Seabrook, J. 1990. *Asking the earth: Farms, forestry and survival in India.* Earthscan, London.

Schuthof, P. 1990. Common wisdom and shared knowledge. M.Sc. thesis. Wageningen Agricultural University, Wageningen, Netherlands.

Thrupp, L.A. 1989. Legitimizing local knowledge: From displacement to empowerment for Third World people. In: *Agriculture and Human Values* 1 (3). Gainesville, USA.

Toledo, V. 1992. Utopia y naturaleza: *El nuevo movimiento ecológico de los campesinos e indigenas de Latino America.* Nueva Sociedad, Caracas.

Van der Ploeg, J.D. 1990. Farmers' knowledge as line of defense. *ILEIA Newsletter* 6 (3). ILEIA, Leusden.

Warren, D.M. and Cashman, K. 1988. Indigenous knowledge for sustainable agriculture and development. IIED Gatekeeper Series No. 10. IIED, London.

Address
Bertus Haverkort, ILEIA, P.O. Box 64, 3830 AB Leusden, Netherlands.

Farmer to Farmer: The Ometepe Project, Nicaragua

Eric Holt–Giménez and Orlando Cruz Mora

Introduction

The degradation of Nicaragua's natural resource base, resulting from uncontrolled development, and the need to develop environmentally sustainable forestry and agricultural systems was documented as early as the 1970s.

Ten years of civil war (1979-89) have since practically destroyed the country's economy and seriously undermined its institutional capacity to respond to environmental and production problems. The land reform implemented by the Sandinista Government (1979-90) successfully redistributed agricultural land to nearly half the Nicaraguan peasantry and established large, publicly owned agro-industrial farms for export crops. However, due to a combination of war, inexperience and underdevelopment, the Ministry of Agriculture was unable to maintain the economic viability of the state farm system and failed to provide a steady flow of credit and agricultural services to farmers. This resulted in a significant decline in productivity throughout the 1980s.

The Nicaraguan Institute for Natural Resources (IRENA) estimates that the country's forests are being destroyed at an alarming rate, 150,000 hectares being lost in 1991 alone. At this rate Nicaragua will lose all of its forests within 20-25 years. Deforestation, and the consequent loss of topsoil from watersheds, has already altered the discharge of rivers and lakes as well as depleting aquifer levels. Climatic zones are also changing, to the detriment of agricultural production.

In conjunction with the national Program for Conservation for Development, and the Action Plan for Forest Resources, IRENA has formulated a policy for the sustainable development of Nicaragua's natural resources. The policy emphasizes the liberalization of the economy, including the promotion of both traditional and new export crops.

There is great concern that this strategy, coupled with the policies of structural adjustment, the poor record of state and regional institutions in working with the *campesinos*, the weak bargaining position of the latter, the mechanisms of debt swaps and other factors will lead to:

° A great reduction in the area set aside as forest reserves
° Induced and short-lived settlement of large areas by *campesinos*
° The continuing displacement of *campesinos* from their plots by ranchers, timber concerns and commercial (export) interests

Because they use diversity to spread risk, *campesino* production systems are potentially well suited to farming under the difficult conditions of the humid tropics, being less destructive of the environment than are plantation and ranching systems, for example. *Campesino* colonizers bring a wealth of farming knowledge with them when they move into a new area. Indeed, experience suggests that, contrary to popular belief, they are capable of generating innovations which significantly reduce migration, and that they, rather than technicians and agronomists, hold the key to successfully transferring innovations among rural populations.

The Campesino a Campesino Movement

Origins

The Campesino a Campesino (Peasant to Peasant) Movement enables *campesinos* to organize locally and nationally to help themselves through peer training in food production and sustainable agriculture. The movement starts with felt needs and envisions a *campesino*-led transformation of Nicaraguan agriculture. Its roots lie deep in the Guatemalan and Mexican experiences of farmer-led development. It is based on the premise that the best way of generating and transferring appropriate technological options is through small-scale, farmer-led experimentation and the direct exchange of knowledge between *campesinos* from different countries, regions, villages and farms.

The Nicaraguan Revolution and agrarian reform of the 1980s thrust *campesinos* into the vanguard of social change, opening national and international communication among them as a class. This had a profoundly emancipating effect on them. For the first time they were able to organize freely on a large scale, forming the Union Nacional de Agricultores y Ganaderos (UNAG). This led directly to the Campesino a Campesino Movement—an organizational leap forward from the externally dominated forms of training, technology generation and transfer organized by non-government organizations (NGOs) which predominated before 1986 to new *campesino*-led initiatives for 'production and protection'.

Between 1986 and 1989, UNAG, in a collaborative pilot project with a Mexican NGO, the Servicio de Desarollo y Paz A.C. (SEDEPAC), organized a series of training visits between Mexican and Nicaraguan *campesinos*. The objective was to train the Nicaraguans in soil and water conservation techniques that the Mexicans had found effective. The Mexicans had been trained originally by Guatemalan *campesinos* under a World Neighbors project at San Martin Jilotepeque, during the 1970s.

During these visits the Nicaraguans learned a process of experimentation and peer training which not only led to a threefold increase in basic grain production but also established a strong Nicaraguan team of *campesino promotores*, keen to pass the message on to others.

Farmer-promoters

The movement is an informal and extensive one. The bulk of its members are simply farmers with a will to experiment and to share with others. The role of these farmer-promoters varies greatly, depending on the communities, institutions and individuals involved.

A common case is the farmer-promoter who experiments on his own farm and works with a local NGO to give workshops attended by other farmers from within or beyond his/her own community. Within the community, training is usually informal. Its organization is supported by the local custom of providing mutual labour to each other's farms.

NGOs support the movement by providing technical assistance, financing for training, field events, visits, and so on. Technical and financial support takes many forms, depending on the NGO involved and the presence and quality of the technicians and of the farmer-promoters.

Salaries for Campesino a Campesino work are a delicate issue. Many promoters are decidedly against being put on the payroll as this would place them at the mercy of the NGO concerned, would limit the number of promoters able to work in the local community and might even have the effect of demoting the promoter, since the status of salaried worker would reduce his/her credibility among farmers. Some programmes employ promoters for 1-3 days per week only, or only when they put on workshops outside their local community. A rule of thumb in the movement is to pay the promoter twice the local farm worker's daily wage: one wage for him/her and the other for someone to work on his/her farm for that day.

After many unsuccessful experiences in the 'pre-selection' of farmer-promoters through patronage or even democratically, the movement has ended up by relying on 'natural selection'. Those *campesinos* become promoters who participate regularly in training events, who implement and willingly share innovations and who demonstrate a capacity to teach others.

Some promoters, for lack of time, ability or desire, do not give workshops. Their radius of dissemination tends to be limited to their neighbours and extended families. Many of them are excellent innovators and are quite willing to receive groups or individuals on their farm for field days or informal training.

The strongest incentive to be a farmer-promoter is success on one's own farm. The prestige gained by being a good farmer and by sharing knowledge with others comes a close second. Third, a sense of owning and contributing to a process vital to the nation's agriculture also provides considerable motivation. Perks also play a part: trips, tools, seeds, credit and even a small salary are used by different projects and programmes to different degrees. Interestingly, the more material incentives are involved, the more specialized and limited in scope and extent the movement appears to be in a given area. The more social and moral its appeal, the broader the movement. At the latter end of the spectrum, the movement relies more on the rotation of promoters and on teams rather than on individuals.

The movement probably needs a bit of everything, but if it is really to be a movement of, by and for the *campesinos*, it must be careful not to specialize, formalize or subsidize too much.

The spread of the movement

Using UNAG as an organizational vehicle for nationwide, grass roots communication, farmer-promoters were able to stage short, practical courses in soil conservation, both in cooperatives and on small family plots. The success of this project was broadly and rapidly communicated among *campesinos* throughout Nicaragua. Dissemination occurred spontaneously, without being included in UNAG's programmed activities. *Campesinos* who attended UNAG rallies and meetings spoke enthusiastically about the increasing yields achieved in the pilot areas of the project. Word spread rapidly, and soon *campesinos* from all over Nicaragua flocked to the pilot villages, usually with the assistance of UNAG or a rural NGO.

Short practical courses, field visits between villages and small-scale experimentation were the core activities of the UNAG project. *Campesinos* from different regions and villages put pressure on NGOs and/or their local UNAG offices for activities with Campesino a Campesino. Training was expanded to include training of agricultural technicians and advisors from other NGOs in the Campesino a Campesino methodology. Soon, through direct farmer-to-farmer contact and support from local NGOs, different courses, activities and materials emerged in nearly all parts of the country.

The Ometepe Project

The steep slopes, fragile soils and dense human population of Ometepe Island, in Lake Nicaragua, make the environment for agriculture a sensitive one. Recently, declining yields, lack of credit and the high prices of chemical inputs have forced the island's *campesinos* to seek alternative models of production. The methods they are now trying are those commonly associated with low-external-input and sustainable agriculture (Haverkort et al, 1991).

The Ometepe Agro-ecological Project, sponsored by UNAG and financed by a Belgian NGO, aims to introduce new technology to make agriculture more sustainable, improving both the environment and the livelihoods of *campesino* families. The technology covers soil and water conservation, soil improvement, the control of insect pests and weeds, and reforestation. It is being introduced using the Campesino a Campesino approach.

The project has completed the first stage of a Farmer-led Experimentation Program (FEP), in which a group of *campesino* experimenters have identified the factors limiting their production, and then designed and implemented experiments to test possible solutions. Through collective evaluation, the *campesinos* have evaluated their results from a technical, methodological and organizational viewpoint. The analysis covers the period from May to December 1992.

In Ometepe, as in many areas of Nicaragua, farmers were organized in credit and service cooperatives and worked land titled under the Sandinista land reform programme. They belonged to UNAG, which provided technical support and facilitated the provision of credit by the Ministry of Agriculture. They also belonged to a national rural consumers' cooperative sponsored by UNAG.

With the economic crisis and the privatization of agricultural credit, cooperatives stopped applying for credit for fear of not being able to make repayments and of losing their land titles as a result. The cooperative's store folded. This left the farmers with little actual cooperative activity, but some experience in organizing for collective action.

Objectives
The objectives of the Ometepe Project are:
- To protect the island's natural resources and to create a new equilibrium between agriculture and the environment.
- To introduce sustainable, environmentally sound production techniques, in order to recover and maintain soil fertility.
- To promote the participation and self-management of the beneficiary group (poor *campesinos*).
- To increase agricultural production and productivity.

Within this framework, the FEP has the following specific technical, methodological and organizational objectives:

Technical
- Identify the critical environmental factors limiting production in the island's Moyogalpa municipality.
- Identify and test *campesino* techniques which overcome or circumvent these limiting factors.

Methodological
- Teach simple techniques which allow:
  - Problem definition
  - Hypothesis formulation
  - Design of experiments
  - Data collection and analysis
  - Communication of results.
- Identify and provide feedback on indigenous knowledge.

Organizational
- Train a group of *campesino* promoters in individual and collective experimentation.
- Establish relations with other *campesino* groups for the purposes of experimentation and promotion.

Methodologies
The FEP follows the methodologies of participatory technology development (PTD) and the Campesino a Campesino Movement.

PTD does not happen in a vacuum, and as such should be viewed only as a tool for revealing and unleashing the local potential to share and improve indigenous knowledge. Local churches, cooperatives, farmers' associations, NGOs and government programmes can all be accessed for inputs to the initial stages of PTD. In some cases they can participate throughout the PTD process. This depends largely on their ability to accommodate and support new initiatives by farmers.

The 26 promoters selected by the Omepete Project to begin the PTD process were contacted through the local UNAG chapter and the local Catholic church. They were almost all members of the cooperative and had previously participated in either church or UNAG credit or marketing projects.

After the first workshops were held the farmers began to organize for group experimentation. At this point they formed new organizational structures and mechanisms to give them more decision-making power within the project. These included a Consulting Committee, the sending of representatives to the project's Executive Committee, and Campesino a Campesino training teams. This had a dynamic effect on the PTD process, attracting more farmers and increasing the participation of those already involved.

The Omepete Project thus provides a typical example of the PTD process at work. Participants have gradually assumed more and more control, not only of the experimentation and training activities, but of the project itself. The PTD steps followed are shown in Table 1 and described in more detail in the sections which follow.

The Participatory Technology Development Process

Step 1: Getting started
The work on Ometepe Island began after various groups of *campesino* producers had participated in events of the Campesino a Campesino Movement off the island. *Campesino* promoters gave the first two workshops, helping us to take our first steps in the PTD process (field survey/problem definition), and motivated the group to discuss the movement and use its methodology.

The initial stage of PTD, which involved reconstructing the agro-ecological history of the island and of farmers' plots, relied heavily on farmers' inputs. A rapid field survey was conducted with 26 producers. Five groups of producers took soil samples and described the conditions of the land worked by a particular cooperative.

It soon became apparent that a major reason for declining yields was erosion and the loss of organic matter in the silty/sandy soils of the island's lee shores. Farmers described the yields obtained in the past under slash-and-burn cultivation, without the benefit of fertilizers or machinery. Soil samples from over-cultivated fields were compared with those from mature virgin forests. The connection between fertility, organic matter and biomass was obvious to farmers. The low cation exchange capacity (CEC) of the island's sandy soils and their volatile organic matter content had led to the rapid loss of fertility when the traditional dibble stick and wooden ploughs had been replaced by tractor-drawn discs. Ironically perhaps, the land reform had imposed a permanent agricultural system upon the migratory, slash-and-burn system of the formerly landless *campesinos*. Without the benefit of the traditional 7- to 12-year rotation, wind and water erosion rapidly degraded the soils, on which farmers continued to slash and burn before planting, even though the lack of vegetation made it difficult to pull enough fuel together to start a fire.

The decline in soil fertility had been compensated for temporarily by increasing inputs of chemical fertilizers, from 64-127 kilogrammes per hectare to as high as 318-386 kilogrammes per hectare within a 7-year period. During this period yields of sesame had fallen drastically, from 1284-1947 kilogrammes per hectare to 778-973 kilogrammes per hectare. Maize had suffered similar setbacks, despite the introduction of improved varieties. Then had come severe economic recession, which had effectively denied small farmers access to inputs. Relying on the island's degraded soils alone, yields had plummeted still further. In a recent dry year, many farmers had not even harvested as much seed as they planted. Farmers concluded that mechanization and heavily subsidized fertilizer application had pushed the island's fragile soils beyond the limits of their endurance.

Besides low fertility, the soils' low water holding capacity and the competition from weeds were recognized as priority problems. In addition, the explosion of pest populations was directly related to the destruction of the island's biological diversity and the introduction of pesticides. The relationship between ecological equilibrium and sustainable production was explored.

Various soil conservation and fertility improvement measures were proposed as remedies to this situation. At the first workshop, the promoter explained how to make compost in only 30 days. At the second workshop, another promoter taught participants how to make an A-frame and to use this to lay out contour ditches, bunds and various other erosion control structures, including terraces. Minimal ploughing in furrows was the recommended cultivation practice to accompany these measures.

Step 2: Identifying useful elements

The selection of options to be tested took place at the third workshop, on Small-scale Experimentation. The importance of pilot testing on the same land, before implementa-

Table 1 Six steps in participatory technological development

Activity	Objective	Methods	Results
1. Getting Started • Worshop No. 1 La Flor, Los Angeles February 1992	• Develop cooperative relationships • Preliminary analysis of the situation • Prepare organic compost	• Promoter visits • First workshop, field inspection/ organic compost • Collective problem definition • Compile secondary data	• Participation of 24 cooperative members • List of problems and their causes • Creation of two compost heaps • Library • Project document
2. Identify useful elements April 1992	• Identify priorities • Identify indigenous and scientific information • Select options to be tested	• Second workshop: Small-scale Experimentation • Third workshop: Soil and Water Conservation • Analysis of secondary data • Typology of farms/producers	• Participants familiarized with CCT methods and promoters • 28 farmers trained in conservation and experimentation practices and techniques • Prioritized list of problems, causes and possible solutions • Selection of the target area (Moyogalpa) • Registration of farms
3. Design of experiments May-July 1992	• Review current experiemental practices • Plan and design experiments • Design evaluation procedures	• Fourth, fifth and sixth workshops: Organization of Experiments • Three regional workshops for experimenters	• 8 communities conduct experiments • 18 test sheets for experiments • List of measurements/observations • Characterization of indigenous themes in humid, dry and hillside environments

tion on an entire plot, was stressed. This would allow the farmer to compare the results of alternatives directly with traditional methods, without endangering the harvest, and to experiment with more than one new option at a time. The participants learned the importance of 'only changing one thing' while leaving others the same (control of variables), and of utilizing plot sizes that allow easy quantification and comparison with other areas and methods (person-days for specific tasks and yields per hectare). They also learned the importance of recording their observations and measurements in notebooks.

Table 1 continued

Activity	Objective	Methods	Results
4. Put into practice August-September 1992	• Implementation of experiments • Measurements, observations, evaluations	• Technician's visits to plots • Experimenters' field days • Observations and measurements by technician and farmers	• Creation of the experiment and test plots • Standardized labour treatments and replicates • Test sheets completed (by technician) • Observation book filled in (by farmer) • First collective evaluation of experiments
5. Share the results October-November 1992	• Communicate basic principles and processes • Provide technical training • Offer tested technologies	• Field visits to other locations • Publication of interviews with experimenters • Workshop: Results • National meeting of farmer experimenters	• Results quantified and analyzed • Evaluation of technology • Distribution of 280 copies of Isleña • National network of farmer experimenters and trainers established
6. Consolidate the PTD process November-December 1992	• Creation of favourable conditions for continuing experimentation and development	• Transect: The Moyogalpa watershed • Half Manzana Plan • Ninth workshop: Technical Refresher	• Problems identified for different river basins • Support for broader range of activities provided • Teams of promoters formed

To overcome the limiting factors they had identified (low organic matter content of soils, erosion, weeds), the farmers selected the following options for experimentation:
- Planting velvet beans (*Mucuna pruriens*) to control weeds (*Imperata cylindrica*) and improve fertility:
    - Mixed with maize (*Zea maiz*)
    - Mixed with plantain (*Musa* sp.)
- Applying organic compost as fertilizer on:

- Maize
- Sesame (Sesamo I)
- Plantain
- Controlling water and wind erosion through:
  - Contour ditches, live barriers, retention ditches
  - Dykes
  - Mulch.

To support the process of selecting useful options, the project's technicians and advisors reviewed the available literature about the island, which describes its small-scale, market-oriented farmers. The typical farmer has a low level of technical training, produces staple grains and is associated with a credit and service cooperative. An analysis of the agro-ecological zones identified in the literature, together with the project's own preliminary agro–ecological studies, allowed the recommendation domain for innovations to be established.

The information from the literature was confirmed through a field survey in which each farm was characterized and a profile made of each *campesino* participating in the FEP. The resulting field guide on each participant served to compile a data base for use in the programme's future evaluation.

Step 3: Design of experiments

While completing the field guide on site with the participants, the project technician also made recommendations regarding the design of the experiments selected by each farmer. The details of each experiment were recorded, including the variables to be observed and measured. The technician distributed notebooks so that the *campesinos* could keep records showing the date, the work conducted on the plot and the observations and measurements made. The technician also prepared a test sheet on which basic data about the farmer, the type and objective of the experiment, its design, size and location were recorded, together with general observations about the topography, soil, vegetation and other site characteristics.

This information provides a basis for identifying problems related to data analysis and comparison across locations.

Step 4: Put into practice

A total of 17 'formal' experiments in eight communities were initiated. A further five producers tested the velvet bean, manure and compost innovations on small plots, without having control plots, controlling other variables or recording their data. In his field visits the technician gave priority to the 17 formal experimenters, but the entire group continued to attend project activities. Problems of drought, pests, damage by animals, fires and theft affected many experiments.

Field days, on which participants visited selected experiments, helped to strengthen the group's morale, as well as to detect problems on the experimental plots. The new knowledge acquired while the experiments were under way was disseminated among participants. Obliged to explain his experimental plot under the critical eye of his inquisitive fellow experimenters, the participating farmer saw the importance of recording data and observations accurately. The *campesinos* showed tremendous positive interest and support. Little by little, they helped each other to evaluate the innovations tested and, in so doing, demystified the process of research.

The technician's visits and the field days gradually revealed that while most experimenters had utilized the 'plot-within-a-plot' design, which allowed variables to be controlled and comparisons to be made, almost all experimenters lacked a control plot per se. None of them had designed the standard, recommended plot of 5 x 20 metres. All had taken measurements for their experimental plots, including the seeds or compost applied and the days of labour invested. However, the measurements sometimes mixed volume with weight, and the same parameters were frequently measured in different units.

During this phase, several visits to areas off the island were made, to see the low-external-input technology already put into practice by other *campesinos* of the movement. Because they were conducting their own experiments, the island's *campesinos* had very specific questions regarding the management of the new technologies. They neither accepted nor rejected at face value the explanations given by their colleagues, but listened attentively. Both the experimenters and the promoters showed considerable interest in these visits and appeared to have been highly motivated by them.

Step 5: Sharing results

As the first harvest approached, the project technician visited every participating farmer, providing advice on harvesting methods and on quantifying the harvest from experimental plots. A particular concern was to avoid mixing the experimental with the normal crop harvest. As harvesting began, the technician measured standard areas (5 x 20 m) within the experimental and larger plots. He checked the results recorded by farmers, changing some and discarding others, but accepting the majority.

To disseminate the FEP's first results, the project technician interviewed the four experimenters with the most accurately quantified results in the areas of fertilizer application and weed control. Using a *testimonio* (story-telling) style supported by basic experimental data, he wrote articles for the *Ometepetl Bulletin*, the monthly newsletter published and distributed on the island by the project.

A workshop was conducted at which the 17 participants of the FEP presented the results of their experiments to everyone else. The results were discussed and analyzed from technical, methodological and organizational viewpoints.

In general, the experiments had confirmed the value of the introduced technologies, which reduced costs, labour inputs and/or risks, and increased production in the majority of cases. The experiments resulted in renewed interest in low-external-input and sustainable agriculture among the participants. The different practices used by different members of the group led to several possible conclusions concerning the cultivation of velvet bean, in terms of planting date and density, association with other crops and positioning of the seed. A lively discussion on the appropriate application levels of organic compost took place, which led to the design of further experiments for the second season. The experimenters allowed their velvet bean experiments to continue growing, so as to obtain seeds for the first planting of 1993.

Methodologically, the group had great difficulty in expressing their results in comprehensible units. While they knew which treatment had obtained the best results, they could not translate their data into quintals per manzana, the most widely used measurement in the country (approximately 0.7 hectare). In each case they had to resort to the separate measurements previously made by the technician, who had anticipated the problem. The impact of this experience was significant: in the round of experiments designed for the second planting (August–November), the percentage of experimenters using standard experimental plots (5 x 20 metres or 10 x 10 metres) increased from zero to 60%.

In terms of organization, the group confirmed its desire to continue being trained as Campesino a Campesino promoters. They also proposed organizing themselves into work teams, so as to implement the techniques they had tested over entire plots.

The advantages of experimenting as a group were epitomized by one participant, who observed: 'I only brought one experience here to share, but I received sixteen others in return!' Undoubtedly, the fact that there were 17 experiments on the same innovations helped the *campesinos* to reach general conclusions about the performance of these innovations and their appropriateness for different farming conditions. The difficulties they had encountered in quantifying their results were partly overcome by testing general observations against the data from the four plots on which reliable results had been obtained.

The results from the first season awoke interest in testing the application of manure and compost to red beans *(Phaseolus vulgaris)* during the second season. However, the late harvest of the previous crop (sesame) meant that existing experimental land was not free in time to be used for new experiments. Four new experimental plots were therefore added during the second season to allow research on this theme.

Step 6: Consolidate the PTD process

The initial results of the FEP were presented by the participating experimenters and the technician at a national symposium of *campesino* experimenters organized by UNAG's Campesino a Campesino Program. Twenty-two experimenters from 12 communities located in different regions of Nicaragua made presentations on the following themes:
- Soil and water conservation
- Velvet bean (fertility and weed control)
- Organic compost
- Natural insecticides
- Crop diversification
- Forestry activities.

The main contribution of the experimenters from Ometepe lay in their ability to communicate their results quantitatively, in terms of quintals produced per manzana and per work-day. The event strengthened the participants' desire to experiment and further helped to evaluate the options tested, as well as playing an important part in disseminating the results to other potential users in the movement, elsewhere in the country.

The 23 producers testing velvet beans have committed themselves to sharing seeds and teaching the techniques they have learnt to five neighbouring farmers each over the coming year. This will expand the number of experimenters to more than 100 for the second year of the project. While it is doubtful that each of these will implement the mechanical conservation measures (due to their large labour requirements), it seems probable that all will plant velvet beans, the demand for which is threatening to outrun supply on the island. Packages of 1-2 pounds of seed, sufficient for experimental plots, will be provided to support the spread of this technology.

Following the FEP's successful initial results, the producers wished to adopt some of the techniques they had tested over their entire plots. However, two problems discouraged them from doing so:
- The high labour requirements of the physical conservation measures meant that they could be introduced only very gradually.
- The traditional practice of allowing cattle to roam and graze freely after the second harvest (December) might harm not only the velvet bean harvest (February) but also the bunds and retention ditches constructed for soil conservation.

To solve these problems, the 'half manzana' plan was formulated, whereby the project undertook to make loans from a rotating fund so that barbed wire for fencing on 0.5 manzana per producer could be purchased and erected. This would provide a protected area on which to introduce the new techniques.

The area of 0.5 manzana was selected because it was small enough for producers to have sufficient labour to implement the physical conservation measures, yet large enough for the resulting yield increases and lower production costs to be sufficient to pay back the cost of the barbed wire after one or two cropping cycles, without placing undue stress on the family economy. In addition, this size of area allowed for sufficiently rapid rotation of the fund to expand the number of farms covered by the local Campesino a Campesino programme.

The plan was to involve 30 producers in 1993. On average, three of the techniques tested in the previous season will be implemented by each producer. In addition, a reforestation plan was incorporated in which each producer specified the need for trees on the half manzana. To meet demand, a nursery with 15,000 plants of nine different species was established.

The 30 producers also participated in a meeting at which the project's Annual Operational Plan was formulated. At this meeting the problems of crop diversification and seed production were discussed, and the next themes for collective experimentation identified.

Conclusions

Of the 30 producers present at project initiation, 18 participated in the FEP—the project's main activity during its first year.

These producers were not experimenting with a package of innovations but rather with a menu of different options from which they could choose and which could be evaluated through a shared experimentation programme. In such a programme the diversity of management techniques and approaches used by a group attempting to solve common problems can be taken advantage of and used to make the research programme more cost-effective. While each producer experimented with only one or two options, nine different options were tested and the results were shared among the entire group in only 1 year.

The great diversity of technical skills and methodologies covered by the FEP would have been very difficult to manage if the FEP had sought to apply the rigorous technical standards normally expected of a research programme. But given the real conditions existing in the countryside—the fact that production is low, that there are two or three basic limiting factors, and that it was necessary to produce recognizable results quickly to sustain enthusiasm in the voluntary programme, this diversity was the FEP's most valuable resource. While some *campesinos* are rigorous in measuring, others are meticulous in observing, while others are more imaginative, capable of inventing possible new solutions. Some are good at sharing and teaching, while others are 'good producers', whose farms serve as an example to others. In the end, all the participants were indispensable to the programme's success.

Another indispensable element was the technician's respectful and diligent support. His objective was not to persuade participants of the value a new technique, but rather to help the group select and evaluate techniques on their own. In this he succeeded admirably.

Finally, the FEP was part of a process in which the *campesinos* took increasing responsibility for directing the entire project. Direction was provided by a Consultative Committee consisting of all the participants. We feel that the PTD approach has been consolidated, both within the project (through the Committee) and beyond it (through the Campesino a Campesino Movement).

The FEP will be maintained in the coming years to tackle other limiting factors and critical environmental problems. So far it has shown itself to be a highly suitable approach for intensifying the search for *campesino*-based solutions to agricultural problems in the fragile environments of Nicaragua.

Reference

Haverkort, B., van der Kamp, J. and Waters-Bayer, A. (eds) 1991. *Joining farmers' experiments: Experiences in participatory technology development*. ILEIA/Intermediate Technology Publications, Leusden and London.

Address
Eric Holt-Giménez and Orlando Cruz Mora, c/o Fundacion Entre Volcanes Aptdo 3893, Telcor Central, Managua, Nicaragua.

Farmers Search for New Ways of Cooperating: Networking in the Friesian Woodlands of the Netherlands

Wim Hiemstra, Fokke Benedictus, René de Bruin and Pieter de Jong

Introduction

Through the centuries, farmers have managed and shaped nature and landscapes. The Friesian Woodlands, part of a northern province of the Netherlands, is perhaps the most important Dutch example of a traditional Western European landscape. In this region, the original landscape of wooded earthen bunds, with tree borders used as fences, has been preserved, thanks to farmers' efforts.

This preservation is remarkable, as Dutch agriculture has undergone increasing industrialization since 1945. Intensification, specialization and economies of scale achieved through labour-saving mechanization have been part of this process. Declining numbers of farmers keep more cows and cultivate more hectares, while at the same time increasing yields. Nowadays, most farmers produce only one or a few products in bulk for the European and world markets. These trends were stimulated by national and European agricultural policies until 1983. The growth in production has brought food security and a more prosperous farming sector, but it has also led to overproduction and harmful side-effects: nature, the environment, the landscape and the economy of rural areas have all suffered and are now under great pressure. Overproduction results in prices for many products that are simply too low. In future the incomes of Western European farmers will be at even greater risk, because their products will have to compete with cheaper ones from Eastern Europe and their enterprises will be less and less subsidized by government.

All this, together with social pressures for stricter environmental controls, makes the future of the Dutch family farm uncertain. Many farms do not appear to be viable in an agricultural system oriented towards exports. This is especially true for farms in the Friesian Woodlands, whose small size and small fields lead to higher production costs. Nevertheless, research in this area (de Bruin and van der Ploeg, 1991) has shown that there are ways of farming that are both economically profitable and sensitive to this region's remarkable traditional landscape. It was found that many farmers (91%) are prepared to maintain the natural environment. In the future, however, they will only be able to do this if society is willing to pay for it, either through subsidies or through market mechanisms.

To explore the possibilities, two farmer networks have been established in the Friesian Woodlands. The farmers feel government policies towards agriculture and the environment as they stand at present are confusing and contradictory, providing them with little incentive, but they continue to search for their own solutions, integrating the twofold need to produce and to preserve. The farmers wish to take responsibility for the landscape, but to share this responsibility through a new relationship with government in which they will have more freedom to develop imaginative solutions at farm level. Long-term contracts (10-25 years) between farmers and government, with financial compensation for managing the landscape and the environment, could be part of this relationship. These contracts would also stimulate employment in the region. Agriculture is the mainstay of the local economy, and making the preservation of the traditional farming landscape more profitable for its custodians is important for increasing both the quality and the economic viability of rural life.

The Friesian Woodlands

The northern Friesian Woodlands are characterized by miniature landscapes with small villages. Within short distances (10 to 30 kilometers), different soil types occur: higher lying sandy soils, changing to clay and peaty soils in depressions. The peaty soils are excavated, and because of the increasing use of fertilizers through the centuries, their flora and fauna have degraded. Yet, compared to other areas in the Netherlands, degradation has progressed much less rapidly. More open areas, where canals divide the land into plots and ensure good drainage, are interspersed with areas that are more wooded, with alders or hawthorns bordering the canals. In addition, earthen bunds approximately 1 metre high and planted with trees (mainly oak and alder) function as fences on the higher sandy soils. Scattered around are small watering holes (20-50 metres across), where cattle can drink.

Agriculture
Around the year 1600 A.D., cropping on poor sandy soils depended mainly on the application of cattle manure to maintain soil fertility. The less suitable peaty soils were communal lands used for grazing sheep and cutting sods for fuel. The crop lands were mainly situated around the village, while the communal pastures were further away. Wooded earthen bunds served to prevent livestock from entering crop lands from the pastures.

Agriculture in the Friesian Woodlands in 1993 consists mainly of cattle keeping and is characterized by its small scale and relatively extensive nature. Few animals are kept indoors. Farms consist of many small pastures with a low stocking density, only 1.5 milk cows per hectare. This compares with up to 3 milk cows per hectare in other sandy, small-scale areas. In these other areas more inputs are needed and a great deal of manure is applied. Ammonia, which is found in cattle manure, is thought to contribute to acid rain.

Networking in the Friesian Woodlands

Because trees and shrubs are sensitive to acidity, their vigour is affected. The more cattle, the more manure, the more acid rain.

Policies for the rural areas
In the Netherlands, agriculture and nature conservation come under the Ministry of Agriculture, whereas the Minstry of Housing, Planning and Environment is responsible for environmental protection. Since 1945 the government has concentrated its agricultural development efforts on those farms able to incorporate modern science-based technology. Farmers who decided to grow, invest, manage more livestock and intensify were termed as having vanguard farms. They were supported through subsidies and the services provided by research and extension. As has now become clear, the larger farms especially have benefited from this support. Of the total subsidies provided by the European Community, 80% go to such farms (SAFE, 1992).

Areas where the factors of production are less than optimal are written off as far as agriculture is concerned. In these areas, nature reserves are planned and agriculture is strictly limited. Restrictions on the activities of farmers are, in theory, compensated by subsidies for land improvement. Farms considered viable are offered the opportunity to transfer to other areas where agriculture is not restricted. However, 'viable' may be differently defined by different people. In recent debates the vanguard farm has once again been taken as the model for sustainable development. The diversity of the country's agricultural sector is seen more as a problem for effective policy implementation than as a source of possible solutions.

For nature conservation, the Ministry of Agriculture has formulated its most important objective as the '...conservation, rehabilitation and development of nature and landscapes'. In this policy plan for nature, a strong plea is made for a 'national ecological network' throughout the Netherlands, connected with major nature conservation areas in Belgium and Germany. This network would consist of core areas, nature development areas and ecological corridors. Nature would be further protected by buffer zones, designed to cushion against external influences. In this way, areas would develop that are physically contiguous and thus flora and fauna would not be isolated and so would have more chance of surviving. The idea behind this plan is good, but there is one disadvantage: farmers would have to give way. In the ecological network, agriculture would face many restrictions. For the benefit of nature some 90,000 hectares of agricultural land would be acquired and turned into areas of high ecological value. Between 1990 and 1997 the agricultural lands to which the policy plan applies would be doubled in extent from 100,000 hectares to 200,000 hectares, which is 10% of all agricultural land. Land once reclaimed from the sea would be turned into swamps and lakes again.

Because of these government policies, production and preservation are becoming less and less easy to reconcile. There are plans to take farm land in the Friesian Woodlands out of production altogether, to recreate the natural environment. In the municipality of Achtkarspelen, about 1500 out of 6000 hectares of farm land would be turned into a nature reserve, with swampy forests. Existing tree stands on field borders would have to be replaced by new ones, about 16 to 17 metres wide. The farmers, however, feel this does not accord with the traditional landscape. They argue that this is an environment that has not been entirely 'natural' for centuries, and that reverting to nature now would mean that a part of Dutch history and culture would disappear. They also argue that remaining farm land would have to be farmed more intensively to achieve the same total output. And this, of course, would do further harm to the environment.

Besides these policies regarding nature conservation, the Ministry of Housing, Planning and Environment has formulated an environmental policy plan. It consists of sweeping measures that show little regard for the diversity of the agricultural sector. The main aim is to resolve environmental problems before the year 2015. One of the objectives is to reduce acid rain to a negligible level. Tree borders around Eastermar are thought to be sensitive to acid rain. For this reason the local government department responsible for implementing the ministry's plan proposed the following measure: when a farmer stops farming, other farms will not be allowed to grow in size. Livestock populations will be kept at their present level and tree borders will be preserved. If it becomes law, this will stop the further economic development of the area.

For a number of farmers, these plans, and their uncertain scientific basis, were the straw that broke the camel's back. The landscape would become a millstone round their necks. Having maintained it in earlier days, they would now be punished for doing so. In areas where nature was to take priority, some farm families would have to move away altogether. In other areas, agriculture would face many restrictions. All this caused farm families great uncertainty over their future on the land. They became even more worried when articles about their area began to appear in the media. Anxiety reached such a pitch that the provincial government countered by issuing its own information leaflet 'to create clarity (amidst) the maze of plans and insecurities'.

The Wageningen Study

In 1990-91, Wageningen Agricultural University conducted a study of the Friesian Woodlands, with special attention to the relationships between agriculture and the management of natural resources. In this study, policy options for the future development of the region were to be listed and evaluated (de Bruin and van der Ploeg, 1991). This study made the tensions we have just described explicit. It recommended that a broader range of options for the management of nature should be considered, based on the diversity of existing farming practices. It also suggested that local farmers should be more actively involved in formulating policies for the region. In this way solutions could be tailor-made to suit a variety of circumstances. Central to the analysis was the idea of 'styles of farming'.

Styles of farming

The research showed that farmers in the Friesian Woodlands have very diverse styles of farming. Based on their locality and on the specific history of the farm and its family (who may have managed the farm for centuries), styles of farming were defined by the researchers as 'the total of related ideas, shared by the farmers, referring to the organization of production and development of the farm' (de Bruin and van der Ploeg, 1991). Different styles of farming were recognizable in different patterns of action and practices (caring for livestock, pasture management, management of tree borders, breeding, etc). The relationship between the two was the study's central concern.

Through interviews with 'area experts', the researchers obtained a first impression of the character of the area and the diversity of its farming. The intensity of milk production (milk yield per cow) and the scale of production (number of cows per labour unit) were used as axes for making a social map of the area (see Figure 1).

Next, 25 farm managers were selected for in-depth interviewing. The selection was based on diversity—farming differently to other farmers. During these interviews, each farmer's ideas on the 'correct' way of farming were sought and compared with his or her actual practices. The proposed map of farming styles was offered for comment. Interviewees were asked to describe the differences between farmers in their own words, to explain these differences—and to give examples by naming names. Farmers considered the styles shown in the map to be accurate descriptions. In their eyes, there is a clear difference between a thrifty farmer and a risk-spreader, between a business-like farmer and a breeder. These four main types represent the broad spectrum, which the researchers broke down further to describe the most important farming styles. They may be summarized in four portraits (see Box 1).

Managing the landscape

Utilization of the landscape is influenced by the opportunities created by nature, but it also differs with each style of farming. Each style pictures an ideal (future) landscape that is best suited to that style. Farmers work towards this ideal, making minor, and sometimes major, adaptations and changes to their practices and their environment as they go along. Defining styles of farming can help to explain how their ideas on managing the landscape differ. Thrifty farmers and risk-spreaders have a strong preference for traditional farm buildings and traditional pastures with a variety of grasses. Business-like farmers and breeders, on the other hand, prefer modern farm buildings and highly productive pastures. Thrifty farmers and, to a lesser extent, breeders prefer to work in a closed, small-scale landscape, whereas business-like farmers tend to prefer large-scale, open landscapes, with fewer wooded earthen bunds and rows of alders, especially when the price of milk is falling.

Each style of farming uses nature and the landscape differently. These differences are reflected in intensity of land use, especially stocking density. Business-like farmers practise high-input/high-output farming, using higher application rates of mineral fertilizer,

Figure 1 Social map of the Friesian Woodlands

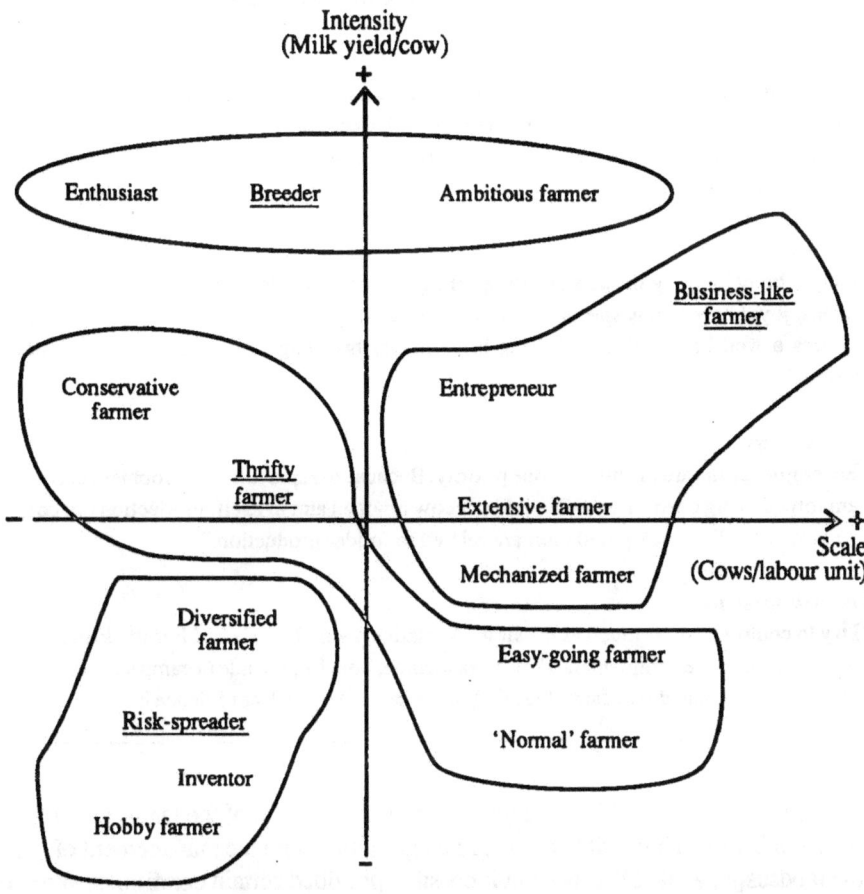

paying considerable attention to the renewal of pastures, to drainage and to leveling. Thrifty farmers, on the other hand, practise low-input/low-output farming, using less mineral fertilizer and hanging back from pasture renewal.

As expected, nutrient losses (the difference between imported and exported minerals) also vary. Business-like farmers and breeders have higher than average losses (336 and 352 kilogrammes of nitrogen per hectare respectively), whereas thrifty farmers cause the least threat to the environment (260 kilogrammes of nitrogen per hectare). The losses of the risk-spreaders were around the average (305 kilogrammes of nitrogen per hectare).

Although tension can be observed between the need for farm development and for preservation of the traditional landscape, the majority of farmers (72%) said that their

> **Box 1. Farmers in the Friesian Woodlands: Four portraits**
>
> *The thrifty farmer*
> "I am not an extreme farmer and do not strive for the highest results. Besides milking, calves and cows for meat production are important as well. Because I have no major debts, I don't have to dig deep. I try to organize my work in such a way that I have time left for other activities as well."
>
> *The breeder*
> "I enjoy breeding very much and caring for my animals, and letting the milk flow is my greatest joy. I have to pay special attention to collecting fodder, because a high milk yield requires a well balanced diet. Selling high-quality breeding cattle provides additional income."
>
> *The business-like farmer*
> "Economic results are my number one priority. Because of rapid farm development earlier, I am forced to dig deep. High milk yield per cow is not an aim in itself. Production per cow and size of the livestock population are related to fodder production."
>
> *The risk-spreader*
> "I try to control costs as much as possible. A relatively small number of livestock and low milk production are complemented by extra income, both from inside (a camp site, making cheese, etc) and outside the farm. Farming this way, I have much confidence in the future."

business interests were not hindered by the small-scale nature of the landscape. Indeed, 33% of them believed it possible to integrate agriculture with the management of nature and the landscape, while 58% thought it possible provided certain conditions were met. The way in which integration can take place again depends on the style of farming.

Under the law as it stands at present, farmers can make agreements with the government, which will pay them to maintain valuable elements in the landscape. These agreements, which are widespread, are valid for at least 7-11 years. Remuneration is such that the costs of both labour and materials are met. Yet although remuneration is reasonable, the farmers are critical, saying the current agreements limit their freedom. The maintenance methods prescribed by government are too rigid: regular pruning, fencing wooded earthen bunds with poles and barbed wire. These rules apply only to mature trees and to earthen bunds; tree borders are not affected. The economic value of the wood is limited, so farmers feel their efforts are not rewarded. Although a majority of farmers (64%, especially thrifty farmers and risk-spreaders) are in favour of integrating agriculture and environmental management, they are unwilling to accept agreements which limit their freedom and which will be difficult to accommodate in their farming practices.

The majority of farmers (55%) suggested that environmental management and maintenance of the landscape should become an economically viable pursuit. Thrifty farmers, risk-spreaders and breeders alike adhered to this opinion, whereas business-like farmers tended to be opposed to it, sometimes strongly so. Investing labour in environmental values did not suit their style of farming, since their high degree of specialization made diversifying their attention almost impossible. In addition, they feared that environmental management would harm their business interests.

Proposal for a new management plan
The researchers suggest that the integration of agriculture and environmental management can be stimulated, starting from the farmers' own point of view. Grants—and the rules and regulations governing them—could be combined in a single fund managed on a regional basis. Farmers would form producers' groups, in which they jointly formulate management plans in line with their own style of farming. These plans would specify the landscape elements to be maintained or created on each farm. There would be a basic package of measures for which farmers would be recompensed on the basis of real costs. On top of that, farmers would be able to choose further measures, the choice being influenced by the farmer's own ideas, resources and style of farming. Thus there would not be a standard solution, but tailor-made options, adjusted to specific farming circumstances.

In the formulation of the plan, direct and indirect costs per hectare would be calculated, up to a certain predetermined ceiling. Both parties—producer and provincial government—would then negotiate. When they agree, a management contract would be drawn up and signed.

Farmers themselves would supervise the implementation of the plan. If implementation were to fall short of what has been agreed, the local government would be able to terminate the contract. Provided sufficient regulatory mechanisms were in place, this would guarantee that farmers meet the commitments they themselves would have proposed.

In this way, farmers could increase the economic viability of their farms if they so desired, while retaining flexibility in their management of the landscape. This approach would cost the government less than contracting the work to labourers, as is currently the practice in some areas. In addition, the financial incentives to farmers would reduce pressure on nature and the environment. Farmers would feel ownership of the scheme, since they would be its principal architects and implementers. Implementation would be more feasible, because farmers' proposals would more closely match their current practices than do government proposals.

Farmers and Researchers Find Each Other

Farmers' associations
When De Bruin and Van der Ploeg presented the findings of their research, discussions on the role of manure in environmental pollution and on the proposed national ecological network were in full swing. The publication of the report coincided with a lecture given locally on the growing importance of farmers' associations, which enable farmers to gain more control over their own future. This led farmers of the Eastermar area to form a working group to further develop the report's ideas. In March 1992, they founded the Vereniging Eastermars Lânsdouwe (VEL, the Eastermar Environmental Association). This was closely followed in November of that year by the foundation of the Vereniging Agrarisch Natuur- en Landschapsbeheer Achtkarspelen (VANLA, the Association for Agricultural Management of Nature and the Landscape in Achtkarspelen). The farmers who formed these associations had been looking for a way forward for a long time. The researchers' study had given them a voice. Now an alternative option was found opposing that of the vanguard farm model. The farmers were able to start working on ways of integrating agriculture and the environment instead of trying to separate them. The researchers helped the Eastermar farmers formulate their ideas into a project proposal, and VEL received a small grant from the Ministry of Housing, Planning and the Environment to develop the proposal further.

The farmers of the Friesian Woodlands decided to form their own associations because it appeared to them that official farmers' organizations were too much concerned with conventional agriculture and with government policy towards this, and not enough with developing their own vision of agriculture and its future.

VEL and VANLA are unique initiatives, mainly because they represent new ways of developing more flexible measures for environmental management that suit the needs of the farmer and the region. These measures form an attractive alternative to top-down prescriptive government policies. Flexibility is not easy for bureaucratic government bodies, but the provincial government appears willing to cooperate and to make these initiatives succeed.

On-farm research
Taeke Hoeksma, together with two sons, manages a dairy farm with 85 milk cows on 55 hectares. Hoeksma is also the secretary and one of the driving forces behind VANLA. He describes himself as a thrifty farmer. Together with 85 others who are members of the association (this is 50% of all farmers in the municipality), Hoeksma has begun conducting on-farm research on integrated management for production and preservation. 'When the experiments lead to clear results', Hoeksma says, 'more farmers will probably join'. Walking around his farm, he points to wooded earthen bunds and tree borders undergoing different treatments. He is also conducting an experiment on his pastures,

trying out different fertilizer treatments. The association plans soon to start experiments on the integrated management of the landscape. This will include the management of canal banks, meadow habitats (for birds), wooded earthen bunds, tree borders and watering holes in a way that is friendly to the environment.

These experiments, begun in 1993, are carried out by the member farms of VANLA. Farmers indicate which experiments they are interested in. The difference between this approach and existing government measures is that, under VANLA, a package of interventions is selected together with the farmer, on the farm, ensuring a sufficiently close fit with existing farming practices for farmers not to feel too restricted. Different levels of remuneration are an essential part of the experiments, including remuneration differentiated according to results. Researchers assist in judging the results during the process of experimentation, and cooperation is sought with relevant government organizations.

Financial support for VANLA is still insecure. The neighbouring association, VEL, already receives support for similar activities. The government wishes to await the outcome of the VEL experiment before supporting other farmers' associations.

The management plans

Later in 1993, VANLA's farmers will organize study meetings to feed experiences from elsewhere into their own farming practices. Based on these meetings and on the initial results of their experiments, they intend to formulate a Plan of Operations. The plan will contain a detailed strategy for integrated management. Funding has been requested to hire someone to maintain existing activities and start up new ones. As Hoeksma observes, 'It is easier to start a farmers' association than to keep it going!'

VEL started somewhat earlier than VANLA. In addition to a grant offered by the Ministry of Housing, Planning and Environment, a temporary assistant of the provincial administration offered to assist VEL in designing its plan. The operations and vision of VEL and VANLA are much the same. VEL takes the current practices of its farmers as its starting point and is attempting to draw up a plan based largely on these. VANLA, on the other hand, is seeking to broaden support through experiments and the development of practical experience before arriving at a Plan of Operations. VEL's plan is expected to take shape in early 1994. By that time, it should be clear whether remuneration for environmental management activities is feasible for both farmers and government.

VEL's plan will take the form of 14 projects, devoted to specific aspects of farm management and the association's role. The Board of VEL, chaired by a farmer, will be responsible for implementation. The first drafts of the projects will be discussed within the Board, after which they will be circulated to other members for comment. In September 1993, the first draft plan (project 6) will be presented to members. This will summarize the results from the pilot projects 1 to 5, and may also include the initial results of projects 7 to 13, which cover longer term and more complex issues. Early in 1994 a final master plan (project 14) should be ready, on the basis of which VEL can draw up contracts with government organizations.

The projects are as follows:

1. Information channels. These will inform members, government organizations and other interested parties of developments. They will consist of a newsletter, extension meetings, a series of audio-visual presentations and a permanent exhibition. The newsletter will be co-financed by the local agricultural bank.

2. Inventory. This consists of individual farmers' proposals for managing the landscape over the longer term, based on their current practices. Initial indications are that on 90% of the farms a young farmer is available to continue farming in the future.

3. Regulations. This is a compilation of all the relevant government rules and regulations. The aim is to assess their effects, and to point out inconsistencies and superfluous rules.

4. Other associations. Links with other associations are being sought, so as to assess their experiences and prevent the repetition of mistakes.

5. Negotiations with the government. The role of farmers in the management of nature reserves (currently in the hands of government) and the maintenance of foot- and cycle paths and/or unpaved sandy roads is under discussion. In 1993 farmers will maintain the unpaved sandy roads for 1 year, after which an evaluation will be made. Levels of remuneration are under negotiation.

6. Draft proposal. Pilot efforts in the above areas will lead to a draft proposal for circulation to a larger number of farmers for discussion and approval.

7. Pollution. Attempts will be made to develop methods for measuring pollution of the area and the role agriculture plays in this. This will be followed by an assessment of what can be done to improve matters.

8. Mineral bookkeeping. To measure the nutrients imported and exported from their farm, farmers will keep 'mineral accounts'. This will be compulsory in 1995, with fines for farmers exceeding an agreed level of losses. A study meeting has already been organized at which ways of reducing losses were discussed by farmers and a researcher.

9. Pollution control. An experiment will be started on two farms to find out how plastics and chemical pesticides can be sensibly and sensitively used, and to explore the more efficient use of energy on the farm.

10. Fauna management. VEL will draw up regulations and a plan for the management of fauna, in consultation with local hunters.

11. Forming associations. What legal status is best for VEL and other farmers' associations?

12. Certification. Is value added to agricultural produce as a result of environmental management? If so, can this be translated into the pricing of agricultural products?

13. Recreation and tourism. What are the possibilities for recreation and tourism in the area?

14. Master Plan. The results of all these projects will be summarized in a Master Plan for presentation to the Minister of Environment.

Small projects make the plan much more manageable. If financing is not sufficient to execute the whole plan, VEL may decide to drop one or more of the projects. Research under projects 2, 3, 4, 12 and 13 is to be carried out by Wageningen Agricultural University.

The added value of environmental management and its reflection in the prices of agricultural products (project 12) is an especially interesting subject for research. When a product is certified as originating from a specific region, the consumer is able to support the farmers of that region by buying it. Higher prices can lead to higher incomes for farmers and increased employment opportunities. A striking example of this may be seen by comparing Friesland with the Italian province of Parma. Both areas have the same level of milk production (1.8 billion litres) and the same level of incomes. But in Friesland there are 6000 jobs in the dairy sector, whereas in Parma there are 30,000. The difference lies in the fact that, in Italy, added value is paid to the farmers instead of to the industrial sector (De Roest, 1990). However, Eastermar has so far lacked sufficient interest and capacity to process the milk it produces.

Plans like those of VANLA and VEL are social as well as technical, formulated from the grass roots, the farmers. Demonstrating that there are alternatives to the dominant scientific-technical model of agricultural development—the 'vanguard' farmer espoused by government—opens up new prospects for other traditionally farmed areas, both in the Netherlands and elsewhere in Europe.

Farming is central to the rural economy of regions such as the Friesian Woodlands. When farmers leave—a trend which set in many years ago but which has been aggravated recently by government policies—schools and shops close, and finally whole villages can be abandoned. This trend must be reversed if the plans of VEL and VANLA are to succeed. In the words of van der Ploeg (1993), 'Such innovations are only possible with sufficient quantity and quality of farm labour. ...In the present situation, we lose those actors which we will be needing so badly in a renewed rural society. A paradoxical situation could evolve, where there are only few farmers left at a time they are needed more than ever.'

Conclusion

Does the future of the Friesian Woodlands lie in a separation of nature and agriculture that flies in the face of tradition, or will the region's farmers succeed in formulating a strategy that successfully integrates the two? Can farmers persuade the government to back their strategy? Will they prove reliable partners in negotiation and implementers of their own plans? Farmers and researchers are working together to find an answer to these questions. All parties—farmers, government and the rest of Dutch society—stand to gain enormously if positive answers can be found. Or are the expectations placed on the projects too high? Only time will tell.

References

De Bruin, R. and van der Ploeg, J.D. 1991. Keeping track: Styles of farming in the Friesian Woodlands (in Dutch). Bos-en Landschapsbouw Friesland-Groningen and Department of Rural Sociology, University of Wageningen, Netherlands.

De Roest, C. 1990. An example of quality: Agriculture, food producing chains and health (in Dutch). In: Ettema, M. and van der Ploeg, J. D. (eds), Tussen bulk en kwaliteit: landbouw, voedselproduktieketens en gezondheid. Van Gorcum, Assen-Maastricht, Netherlands.

Van der Ploeg, J.D. 1993. Styles of farming: An introductory note on concepts and methodology. In: Haan, H. and Van der Ploeg, J. D. (eds), Endogenous regional development: Theory, methods and practice. EC, GCVI Agrimed Series, Brussels (forthcoming).

SAFE 1992. *What future?* Sustainable Agriculture, Food and Environment Alliance, London, UK.

Addresses

Wim Hiemstra, ILEIA, P.O. Box 64, 3830 AB Leusden, Netherlands
Fokke Benedictus, Seadwei 17, 9261 XM Eastermar, Netherlands
Pieter de Jong, Malewei 1, 9261 XP Eastermar, Netherlands
René de Bruin, Wageningen Agricultural University, Dept. of Rural Sociology, P.O. Box 8130, 6700 EW Wageningen, Netherlands.

Fokke Benedictus and Pieter de Jong are farmers and initiators of VEL.

Building a Grass Roots Institution to Implement Low-external-input and Sustainable Agriculture: The Stewardship Farming Experience

Ron Kroese and Cornelia Butler Flora

Introduction

This paper describes how the Land Stewardship Project (LSP), through its Stewardship Farming Program, triggered the foundation of a farmers' network in the Midwestern United States, called the Sustainable Farming Association of Minnesota. It discusses the roots and evolution of the Stewardship Farming Program and its current structure. The process by which the programme was organized is of particular interest, as it is based on the social action approach to community organizing, but uniquely modified to fit rural communities.

A Brief History of the Land Stewardship Project

The non-profit Land Stewardship Project began in 1982 as an organization dedicated to fostering an ethic of stewardship toward farmland in the Midwestern USA. It grew out of a rural humanities education programme known as the American Farm Project (AFP).

Roots in the National Farmers' Union

The AFP had been launched in 1978 by National Farmers Union vice-president Victor Ray. It was funded by a grant from the National Endowment for the Humanities. Compared to politically powerful and more conservative competitors such as the Farm Bureau, the National Farmers' Union is one of the more progressive general interest farmers' organizations in the United States. Under its leader's vision and guidance the AFP was designed as a 3-year leadership development project, in which outstanding young family-farm couples from each of the 23 states in which the National Farmers Union was active were selected to take part.

During the project the farm couples were involved in workshops and activities on four major themes—land, economics, people and community, and rural image. Several of them have emerged as leaders in the Farmers' Union during the past decade. A number of talented rural writers and activists also played a leading part in developing the project.

In 1981 Ron Kroese was hired to direct the project for what proved to be its third and final year. (In 1982, the AFP ended when the newly elected Reagan administration declined to refund it.) Disappointed by the lack of federal funding but buoyed by the enthusiasm of the farm couples, Ray and Kroese decided to form an independent organization that would focus on environmental issues.

Where agriculture meets environmentalism

In the concept of land stewardship, Ray saw the meeting of environmentalism and agriculture. Ray, who served as LSP's first Board chair and remains on its Board of Directors today, pointed out that while farmers and environmentalists often find themselves at odds, especially on public policy issues, most farmers personally identify with the concept of land stewardship, agreeing that it is the responsibility of farmers and landowners to strive to leave the land in as good or better condition than when it was acquired. That concept formed the vision and mission of the LSP.

Ray and Kroese were encouraged in their plans by the fact that, as the decade of the 1980s began, several of the country's major churches issued proclamations stating that the ethical use and distribution of land ought to be included in contemporary Christian social teaching. For example, both the Roman Catholic church with its 'Strangers and Guests' statement on land stewardship, which was signed by 71 bishops in 12 Midwestern states, and the American Lutheran Church's statement, 'The Land: God's Giving, Our Caring', called for a fundamental re-examination of people's relationship to the land.

Having secured a couple of modest grants from private foundations, LSP's three-person staff focused their initial efforts on organizing farmland ethics discussions in counties in Minnesota and neighbouring states that had been identified by the US Soil Conservation Service as having extremely high levels of soil erosion. While erosion from wind and runoff had historically been high in these counties, the problem had been aggravated by federal farm policies of the 1970s that had pushed farmers to expand production and brought some 60 million acres of new farmland, much of it highly erodible, into row-crop and small-grains production. With the cooperation of county soil district officials, LSP staff organized meetings in church basements and public meeting rooms, drawing attention to the local erosion problem and facilitating discussions on reasons for the problem and possible solutions.

By 1985 three facts had become clear to the LSP Board and staff:
- Non-farmers, as well as farmers, needed to be concerned about soil and water degradation if effective public policies were to be implemented to deal with the problem.
- The huge decline in land values of the early to mid-1980s, combined with poor commodity prices, were making it difficult, if not impossible, for farmers to be good stewards even if they were motivated by an environmental ethic.
- Ground and surface water contamination from the misuse and overuse of chemical fertilizers was emerging as a problem at least as significant as soil erosion.

LSP's cultural programmes

To deal with the first of these facts, LSP created several cultural programmes for general audiences. Among them were a one-actress play titled *Planting in the Dust*, about a young farm woman's struggles on a contemporary family farm, a puppet show for children about the role human beings can play in restoring the earth, a one-man 'Music of the Land' programme that featured songs and sing-alongs about farming and care of the land. These programmes, designed for both urban and rural audiences, continue to be widely in demand. With support from the respective state humanities committees, LSP now has an actress capable of presenting *Planting in the Dust* in Minnesota, Iowa, North and South Dakota, Nebraska, Kansas and Illinois. By mid-1991 the show had been presented in more than 400 churches, schools and meeting halls in the Midwest and Canada.

Farmland Investor Accountability Project

LSP's first organized response to the US farm crisis of the mid-1980s was the formation of its Farmland Investor Accountability Project (FIAP). Here again, although the formation of the project was precipitated by economic conditions, concern about the stewardship of farmland was its main focus. Through a study conducted with the University of Minnesota, LSP revealed that by 1986 the nation's 12 major insurance companies held 5.2 million acres of US farmland. About 80% of this had been acquired through foreclosures on loans to farmers who had become unable to repay their debts when land prices crashed in the early 1980s. Most of the farmland was in the Midwest.

LSP became immersed in this issue when constituents, who had become involved with LSP through its land ethics meetings, began notifying staff about the poor farming practices and lack of concern for conservation on many of these farms. Through a combination of tactics and activities, including demonstrations at insurance company headquarters, prayer services on damaged farms and a hearing in 1987 before a US Senate agriculture subcommittee, LSP drew public attention to the severity of the problem. The campaign subsequently led most of the companies to improve their farm management practices and to include conservation clauses in their farmland rental agreements. The campaign also set the stage for LSP's subsequent efforts to change public policies that discourage sustainable farming, as well as its ongoing efforts to develop public and private initiatives that enable young people to get started in farming.

Model County Program

Concern about the pollution of water, particularly in the karst region of southeastern Minnesota, led LSP staff and board to decide to deal not only with the values issues of stewardship-based farming, but with the practical aspects as well. Initially, staff focused most of their attention on one county, Winona, where they carried out a multi-faceted public education programme including workshops for farmers on environmentally sound farming practices, tree plantings on fragile hillsides with fifth and sixth grade students, and a community organic garden. In addition, the staff coordinated a grass roots campaign to get a soil-loss limits ordinance included in the county's revamped land use plan. After 4 years of diligent work by staff and volunteers, the Winona County Commission passed a measure that made soil erosion in excess of 15 tons per acre per year an offence, making Winona one of a handful of counties in the nation where excessive soil erosion is a crime.

Joining forces for public policy reform
As the LSP evolved during the 1980s, it also became clear that farmers face another formidable impediment besides the lack of information and cultural support for sustainable agriculture. Government policies encourage specialization and overproduction and the consequent misuse of agro-chemicals and degradation of soil and water resources. Those working with LSP saw that, while it was important to encourage good stewardship at farm level, overwhelming economic considerations, driven by United States Department of Agriculture policies, virtually dictated what crops and livestock farmers raised and how they were raised. As farmers sometimes put it, 'You need to farm the government way to survive'.

To deal with this situation, LSP staff joined with 20 other primarily Midwestern, farm, food and environmental groups to form the Sustainable Agriculture Working Group (SAWG). The many groups from which the larger new group was formed varied tremendously in size. Some were quite recently formed, others had been going for more than a decade, like LSP. Some groups were politically oriented, others stayed away from policy work and focused only on farmer networking and on-farm research. Groups in some states operated independently, while in others they had formed partnerships with state government and/or public university efforts to promote alternative farming approaches. Increasingly, public university and government-supported researchers were forming partnerships with NGO groups around the country to do on-farm research.

One of the things all these groups had in common was that they were too small in memberships and budgets to influence federal policy in Washington D.C., whereas the chemical and agribusinesses poured millions of dollars each year into lobbying lawmakers. SAWG began when several of the groups from Midwestern states (Iowa, Nebraska, Minnesota, Wisconsin, North Dakota, South Dakota, Illinois, Missouri, Kansas and Indiana) joined together in 1988 to form a coalition. Together they were able to develop policy recommendations (with input from farmers) and set up a two-person office in Washington D.C. whose mission it is to bring their ideas into the federal legislative process. It should be noted that the member groups in SAWG are concerned not only with environmental issues but also with improving the business opportunities for small- and medium-scale family farms. Most share the belief that the USA needs more farmers to practise environmentally sound agriculture. The number of farmers continues to decline (an estimated 300,000 went out of business in the 1980s), and the average age of farmers in the USA is now over 50; very few young people can afford to get started in the current economic climate. This is a major problem, and an ongoing point of debate is whether or not big, capital-intensive farms will be able to make the shift to biologically based farming approaches. Funding for SAWG comes primarily from grants from private foundations concerned to improve the environmental quality of USA agriculture.

In the USA federal farm legislation is formally reviewed and rewritten every 5 years. In the 1990 Farm Bill debate, SAWG succeeded in bringing sustainable agriculture issues to the heart of the discussions, but achieved only modest improvements in policy. Efforts are now under way to make SAWG national in scope, so farmers from all over the country will be represented, and to broaden the membership to include more environmental groups, such as the Sierra Club and the National Wildlife Federation. Things look very promising on the latter front, since more and more environmental organizations see increasing the number of farmers practising low-external-input and sustainable agriculture as crucial to cleaning up the environment. SAWG leaders are cautiously optimistic that the 1995 Farm Bill will bring about further improvements, since more support is anticipated from the Clinton administration. The Bush administration for the most part opposed SAWG's recommendations and supported policies favoured by the chemical companies and established conventional farming interests.

Stewardship Farming Program
By the mid-1980s it had also become clear to LSP staff and Board members that if a land ethic were really to take hold, practical examples of success on the land were needed, as well as changes in attitudes and policies. In 1987 LSP staff began a five-county project that became known as the Stewardship Farming Program (SFP). This programme has created a number of farmer-based activities and events that (1) instill an awareness of the need for a more sustainable agriculture, (2) work on-farm and in groups to identify constraints to sustainability, (3) experiment with more sustainable practices, (4) demonstrate these practices to their neighbours and to the agricultural establishment, (5) identify structural barriers to adopting more sustainable practices, and (6) work to remove those barriers, particularly at the local and state levels.

The programme organized farm families into peer-support, information-sharing chapters of what became known as the Sustainable Farming Association (SFA). The chapters were designed to enable farmers to experiment with alternative farming practices on their own farms and at their own pace, and to facilitate the exchange of information with nearby farmers. While the chapters are governed by their farmer-member Boards, LSP staff continue to contract with the chapters for staff support as needed.

Sustainable Farming Association
Buoyed by the enthusiasm from farmers for these networks, LSP went on to organize three additional chapters in other regions of Minnesota, funded by state government funds and federal research grants for low-input sustainable agriculture, with some additional support from private foundations. There are now six chapters in the SFA of Minnesota, with a total membership of 500. LSP organizes the chapters and provides staff support until local leaders emerge and the chapter becomes self-sustaining through membership

dues and other fund-raising projects. The chapters are linked to one another through a coordinating board consisting of one or two members from each chapter. Besides meetings on farms throughout the year, each chapter holds an official annual general meeting, while the statewide association holds a larger annual gathering to which chapter members from around the state are invited.

In 1990, the statewide association, officially known as the Sustainable Farming Association of Minnesota, became an independent tax-exempt organization. As explained earlier, although the SFA is no longer formally linked to the LSP, it continues to contract with the LSP for staff support and help in organization and recruitment. Requests from Minnesota farmers for the formation of SFA chapters continue to come in to LSP, and long-range plans call for the formation of as many as 10 new chapters in farming regions of the state by the mid-1990s.

In working with local farmers to help them organize for sustainable agriculture, the LSP has introduced an important innovation to improve the stewardship of the nation's natural resources. Although environmentally grounded, the activity of the SFP is quite different from state and national level environmental movements, in that it demands changes in both individual and collective behaviour on a daily basis. The process of organization used should be further analyzed and documented, as it is effective in involving growing numbers of farmers in sustainable agriculture. It is based on the Alinsky model of participatory social change, but adapted to rural settings where overt conflict is less socially acceptable. The late Saul Alinsky was an influential advocate for the rights of the poor in Chicago in the 1950s and 1960s. He developed an effective, aggressive approach to organizing poor people so as to improve housing, the availability of credit, and so on. Some of the staff of LSP had been trained by protégés of Alinsky. They had skills in bringing people together, in facilitating meetings that foster participation, and in encouraging and building grass roots leadership. They knew how to persuade people to come together to create democratic, participatory organizations that meet their needs for information and provide ongoing support and camaraderie.

The processes that have successfully empowered these people to act positively to improve their environment are extremely important and innovative. The philosophy and approach are similar to those espoused by Chambers in his Farmer First methodology, although interestingly none of the LSP staff became familiar with Chambers' ideas until well after the SFP had begun.

How it Works

Goals and objectives
The goal of the SFP is to improve the environmental quality of agriculture by increasing the number of farmers who practise low-chemical-input, conservation-based farming approaches.

The SFP promotes the on-farm development and adaptation of agricultural practices basic to the sustainability of farmland, wildlife and farm families in the rural community. Key components include:
- Public education programmes on economical and environmentally sound farming alternatives.
- Targeted on-farm research assistance to volunteer farm families interested in experimenting with and demonstrating sustainable approaches with technical guidance from SFP staff.
- The development of regional chapters of the SFA.
- A farmer-based effort aimed at encouraging increased public research on environmentally sound farming approaches by fostering cooperation between farmers and land-grant university and extension researchers.

The organizational process

The process of organizing at the local or regional level is a time-consuming one. Local interest and outside funding are the two necessary basic ingredients required to launch a full-time organizational effort and support for volunteers. The mix of local and outside involvement varies at all sites where the SFP is located, but both are present in all cases.

Once a site has been decided on, based on local need, local demand (which is often different from need), and potential for institutional support, staff are chosen by the LSP through careful screening for organizing ability, commitment and knowledge of sustainable agriculture and the environment. Once the staff are in place, an Advisory Board is formed. LSP staff chooses local people known for their concern for the environment and their links within the community.

Together the Board and staff put on a series of informational meetings. Initial meetings are usually held in a church hall and often begin with a performance of one of LSP's plays or a presentation by a musician, since these cultural programmes have proved to be an engaging, non-threatening means of introducing controversial issues. Using existing networks and building on the interest shown at the meetings, staff ask farmers to volunteer to participate in on-farm experimentation for sustainable agriculture. Participation in on-farm research is seen as a key element in building interest and involvement. Criteria for selection include the reputation of the farmer, previous attempts to experiment with sustainable agriculture, and an assessment of their motivation for participation. Farmers desperate to try anything to save the farm are usually excluded. Each farmer chosen is known by the Advisory Board, whose members look for farm families viewed as respectable by the community, although not excluding individuals regarded as eccentric but hardworking. These farmer-experimenters then form the core of a research and outreach group that evaluates potentially more sustainable practices, shares them with their neighbours through field days and discussions, and provides a support group for other farmers interested in trying new techniques and exploring the possibilities of reducing their negative impact on soil, water and wildlife.

At this point, a chapter of the SFA is formed, expanding the formal links among farm families concerned to change their practices. The regional chapters also generate additional informal local groups through a house meeting format, whereby clusters of neighbours interested in sustainable farming meet with a member of the SFA and an SFP staff member to begin discussions, then continue meeting as an informal support group at the local level to share information and evaluate each others' efforts to be better stewards of the land.

Most participants recognize the potential function and worth of the SFA as a mechanism for achieving their goals. Farmer members do not generally proselytize to increase their numbers, as they feel they are still working on the technology needed.

Until recently, regional chapters have been formed in areas where the SFP was located. However, several new chapters are currently being formed solely by volunteers in other areas of the state. Organization nonetheless makes additional demands on existing SFP staff as there is a continuing need for help in networking, knowledge generation and knowledge sharing.

Most farmers involved in the SFP have gone on to join the SFA. However, they express concern about their ability to function as an independent association, remaining—as they see it—highly dependent on LSP staff. A major issue confronting the association is how to maintain activities either without paid staff or by raising funds to hire staff.

Who particpates?

Participating farmers in Lewiston and Montevideo counties have long established roots in the local rural community or in agriculture elsewhere. Participants are not the stereotype ageing hippie back-to-the-landers, hoping somehow to maintain the spirit of the sixties in the bucolic setting of rural America. A number of the male farmers in both counties are people who have been off the farm for a while, often in city jobs that allowed them to finance their farms, since they did not inherit them. But they are all (except for some of the women) of farm backgrounds—born and reared on a farm.

Two areas with long-term projects, Southeast and West Central Minnesota, have both suffered extreme environmental degradation. Southeast Minnesota suffered some of the worst abuses after the original settlement. Winona got its wealth originally from logging—clear cutting—which left the land extremely vulnerable. The settlers adopted the same approach in their farming practices. As a result, some of the country's earliest conservation efforts began in Southeast Minnesota. In the West Central region, the blowing dust of 1992 made it clear to many that drastic measures were needed to change the way the land was farmed. The participating farmers can *see* their cause each day, as they reaffirm their commitment to the land.

The role of farmer-managed experiments and demonstrations

Demonstrations, which are site-specific, are mounted by individual farmers attempting to solve particular problems on their farms. Other farmers can learn from them, and some extension agents are passing on the results of these demonstrations to other farmers. The data that are gathered cannot be aggregated for cross-farm comparisons, however. This is not as problematic for the farmers as it is for academics!

On-farm research is organized by the SFP, with oversight by experienced on-farm researchers. The results are presented to the farmers by the LSP staff. This legitimation and documentation of their efforts has important organizational as well as scientific impacts. On-farm research can never be as rigorous as experiments on the research station. This is offset by the tailoring of such research to the needs of the individual farmer and to the role of the technical innovation within the local farming system. Despite flaws, the documentation of results promotes the further dissemination of appropriate technology through farmer-to-farmer contacts and the efforts of the Cooperative Extension Service.

Participating farmers are most enthusiastic about their experiments. They question each other in detail, analyze what they have done, explain why it does or does not work. The lack of competitiveness as people share research results is particularly noteworthy— quite in contrast to the usual coffee-shop talk of whose yield is highest, whose row is straightest and who has the fewest weeds.

Besides informational meetings and farm visits, other mechanisms to share the knowledge generated are being developed. A package called 'Farming for the Future' includes booklets and videos on farming practices, based on programme experience to date. Based primarily on work in Lewiston but also in Montevideo, four videos with accompanying printed materials are currently in preparation. The booklets are on mechanical weed control, controlled grazing, nitrogen management and on-farm composting. They are well written, emphasizing how farmers adapt specific practices on their own farms. They have short, illustrative case studies as well as clear explanations of the science involved. The videos are excellent visually and convey a great deal of information in an entertaining format.

Impact

On individuals
Despite heroic efforts by SFP staff to document the results of on-farm experiments and demonstrations, it is still too soon to quantify the programme's environmental benefits. However, the social impacts are impressive. Participation in the on-farm research and demonstrations has an amazing impact on the individuals involved, according to their testimony. All had been innovating before participating in the programme, but all had felt very alone in doing so. Established sources of information, including the Cooperative Extension Service, had not met their needs. At times each had felt labelled as some sort of nut or deviant for trying to implement more sustainable practices, particularly since not everything attempted was feasible or profitable. Yet driven by a land ethic, they had

Some sort of nut or deviant...

constantly experimented, constantly sought out a wide variety of information sources on their own as they attempted to create sustainable farms which met their ethical as well as their economic needs. Once they were brought together, they realized that they were not alone, they were not so odd, and they could learn from each other. Their new-found ability to set up parallel experiments, to share the results, to go over details, and to talk about their experiences in public is perhaps the most important result of 3 years of on-farm experimentation.

An important part of the impact of the LSP is that it fosters a highly participatory ethic that involves challenging the established hierarchy of expertise and changing learner/teacher roles in the creation and dissemination of agricultural technology.

Some of the personality changes experienced by project participants—from shy and reticent spectators to outspoken leaders able to present and to take charge—are astounding. The empowering nature of the project was mentioned particularly by the Advisory Board in Montevideo. The strategy of the LSP staff is to step back and let the farmers make the decisions. Initiation of action may take longer, but its continuity is ensured by farmer ownership of the process.

On the community

The participating farmers relate many changes in the attitudes and behaviour of others towards sustainable agriculture. Fertilizer dealers now ask what was on the land in previous years in order to determine requirements. Extension agents include the technologies developed by SFP farmers in the alternatives they offer, and Land Grant researchers participate in field days sponsored by the LSP. Most important, an environmentally aware approach to farming has now been legitimized in the community to the point at which it can be freely discussed at church or in the coffee shop.

One of the most remarkable accomplishments of the SFP in the past few years is the change in the attitude of the extension services towards it. Initially suspicious of each other, the SFP and extension have increased their collaboration over the years, particularly in the West Central region. This is preparing the way for a useful partnership in both research and extension in the future.

Conclusions

The power of the project is empowerment. Through the SFP farmers discover that they are in charge and that they have choices. There is not just one way to farm in order to be profitable and to stay in business.

As a result their practices are changing. Alternatives to resource-depleting farming practices are being developed in the context of the whole farm. A new concern for community participation is emerging.

The LSP has developed an important institutional approach for enhancing low-external-input and sustainable agriculture. Initially part of a national general farmers' organization, it has been reformulated and regrouped as a multifaceted organization able to address practices and policy at the local, state and national levels. Of particular importance is the use of community organizing tactics to build grass roots organizations which are the basis for changing both the attitudes and the practices of farmers anxious to be responsible stewards of natural resources yet remain profitable in an increasingly difficult economic environment. The fact that the SFA and the SFP are not proselytizing organizations, but instead seek simply to unite like-minded individuals, is crucial to their potential longevity and success.

Address
Ron Kroese: Land Stewardship Programme, 14758 Ostlund Trail N., Marine on St. Croix, Minnesota 55047, USA.
Cornelia Butler Flora: Virginia Polytechnic Institute and State University, Department of Sociology, College of Arts and Sciences, Blacksburg, Virginia 24061-0137, USA.

Supporting Indigenous Knowledge for Sustainable Rural Development in Bolivia: The Case of AGRUCO

Stephan Rist

Introduction

After 500 years of colonialism and development, the once intact ecosystems of the Andes have deteriorated markedly. The Green Revolution did little for their ecology. It is time to incorporate an approach oriented towards low-external-input and sustainable agriculture (LEISA) into national and international agricultural development plans.

When the Agro-ecology Programme of the University of Cochabamba (AGRUCO) was planned and launched in 1985, the original idea was to facilitate the transfer of European approaches to so-called 'organic' agriculture to Bolivia. The main objective was to stimulate the dissemination of organic agriculture practices and technologies among and by the institutions involved in rural extension work, both government and non-government.

In this article I will show how, by forcing us to adjust to the reality of the Bolivian context, our experiences have led us away from a transfer of technology (TOT) approach to a participatory technology development (PTD) concept. After an abortive effort to introduce western models of organic agriculture we learned to appreciate Andean agro-ecology and cosmovision. We gradually shifted the emphasis of our work towards re-evaluating and strengthening local Andean technologies and traditional forms of social organization.

Encouraged by the results, we then embarked on PTD and networking at different levels, building on indigenous networks.

The Project is Launched

A conventional start

The project began with conventional scientific research on topics such as the preparation of compost, the improvement of local manure, the production of organic fertilizer from slaughter-house wastes and the use of local rock phosphates.

This first stage can be characterized as the development of an 'alternative technology package', emphasizing the analysis and improvement of biological processes. Technology was developed independently of the rural communities for whom it was intended. Research was conducted at the research station, where scientists sought to improve soil fertility with new crop rotations, mixed cropping, organic fertilizers, biological control of plant diseases, and so on. The package, or components of it, was then tested on farmers' fields. Here the basic unit of analysis was the single plot. The roles of the researcher and extension worker were defined as to produce the new 'technology package' on the research station and then to extend it to the *campesinos*. The role of the rural community was to receive the new package and use it.

The evaluation of the results and impact of this work after 2 years led to a paradoxical conclusion. The new technology package could improve soil fertility and the growth of plants and animals. The results were positive, both on the research station and on farmers' plots. However, in spite of these favourable results, farmers showed little or no interest in the new technology. Likewise, the interest of other institutions was also very low.

A secondary objective of our research had been to acquire a better understanding of the local management of natural resources. This part of the research had been conducted by students of the University of Cochabamba. This was an excellent opportunity for them to get in touch with the realities of rural life—an important experience not catered for by other curriculum activities.

This research gave us the explanation of our paradox. We became aware of the fact that the small-scale farm family is not concerned only with biological processes but must also balance these with economic, social, cultural and personal factors.

Modifying the approach
These results showed us that we had to adopt a more holistic approach. We expanded the unit of analysis from the single plot to the whole complex of production within its socio-economic and cultural context.

We left the research station. This decision was based on the conditions of the Andean ecosystem, which is characterized by a vast range of altitudes within a relatively short distance and an enormous variety of ecological conditions. One and the same family cultivates several different soils, in different climates, using different crop varieties. This makes the transfer of results from the research station practically impossible, since even in the best of cases new technology meets the requirements of only one of the many environments managed by the farmer.

For this reason we transferred the technology development process to the setting of the rural community, changing at one stroke the research method and the roles of the researcher, the extension worker and the farmers and their families.

The research methodology had to become more participatory, including the community in the definition, planning and implementation of experiments. As a consequence, the researcher expanded his functions to overlap with those of the extension worker, while the latter no longer saw himself as merely the servant of the researcher. It was no longer necessary to extend technology from the research station to the community, because the technology was developed and validated in the community itself, with broad participation by community members. This would lead to a kind of auto-extension by the *campesinos*, if they were interested in the technology.

Thanks to this new approach the relationship between the community and the project was transformed from a vertical one to a much more horizontal one, establishing the basis for PTD. This horizontal relationship with the communitiy also implied the permanent integration of the project's professionals into rural life.

To our surprise, over the course of a further 2 years we discovered that virtually all the basic technical elements of sustainable agriculture—crop rotation, mixed cropping, use of organic fertilizers, biological control of pests and diseases, soil conservation, conservation of genetic resources, protection of wild plants and animals, and so on—were already present and used in nearly all the traditional communities. These traditional technologies reflected a knowledge of all that science has told us about the different approaches to sustainable agriculture in Europe. In traditional Andean systems we found not only the common concepts of organic farming but also elements of biodynamic agriculture. We also found a broad knowledge of the influence of the planets and stars on plant and animal production.

These findings led us to a difficult question: if the knowledge system for sustainable agriculture is present, then why are there so many ecological problems in Andean countries, resulting in the widespread deterioration of parts of the Andean ecosystem?

The immediate answer to this question is that, although traditional knowledge and technologies are still present, their use, like the landscape itself, is being steadily eroded.

Andean Agro-ecology and Cosmology

The Andeans perceive the world in the context of a whole cosmos of which every part is alive. This Andean cosmovision expresses itself in many rituals, which are often carried out before the *campesinos* begin a task such as ploughing, sowing or harvesting. It is also the philosophical basis for their technology development. This we discovered through our conversations with farmers, in the following way:

- The mixed cropping of different varieties of potatoes is a result of the following logic: the local potato varieties are classified as 'male' and 'female'. Male and female varieties combined in the same plot complement each other, thus giving a good yield. Harmony, conceived as the complementarity of opposites, is one of the constituent elements of the Andean world outlook.
- Talking about how he managed his potato crop, a farmer in the community of Rodeo said: 'There is a time for sowing, and a time for cropping; it is necessary to work with the earth, to learn from the moon when to sow, to become a friend of the clouds to obtain a

good rainfall'. This brought us face to face with how farmers define the best moment to begin sowing—they consider the phases of the moon and observe the clouds from August until the beginning of the rainy season (October to November).

These experiences made us modify our approach once again. We started placing more emphasis on understanding the world view of the farmers. The erosion of indigenous knowledge was recognized as one of the main factors responsible for the decreasing use of traditional technologies.

Our next step, then, was to conduct research on the Andean cosmovision. Over the past few years this vision has received increasing attention from anthropologists and sociologists. Their work was of great help to us, since we were able to integrate their findings with our own PTD approach. One of the best accounts is that of van den Berg (1990), who explains how the Andeans see themselves (individuals, family and community) as part of a whole and in constant interaction with the other two components of this whole, which are nature and spiritual beings. These form the supra-human society (Figure 1).

In the Andean world view these three components are inseparable, since they are in constant and dynamic interaction and penetration with each other.

None of the three components is superior. Rather, they exist in a dynamic equilibrium. For the Western mind the mutual penetration or interaction of the human with the natural is not hard to understand. But to comprehend the Andean world view it is necessary to include the transcendental component as well—rather more difficult for the average Westerner. Here are two examples:

• Practically all illnesses of human beings are caused not only by nature but by a negative encounter with the transcendental world. An illness or disease cannot therefore be cured only by taking a medicinal plant (despite its scientifically proven effects). Along with the medical treatment a ritual is needed, directed at this transcendental world.

• Every year on Carnival Tuesday the *campesinos* go to their fields to celebrate their crops, just then in full growth, with a ritual called *challa*. This ritual is held in honour of the Ispalla spirit: 'Ispalla is the soul and spirit of all food... It is Ispalla who gives life to human beings, animals and plants'. A good crop, qualitatively and quantitatively, depends not only on appropriate technology but also on the accompanying rituals.

For our own work the most important thing to realize was that a new technology has a chance of being adopted only if a certain conformity or equilibrium exists between the three components of the cosmovision. If one of the components is out of balance or is not taken into consideration, traditional technology disappears and new technology is not adopted.

The objective of the AGRUCO team was now to identify the factors causing the erosion of indigenous knowledge and technologies. The principal factors we found are the following:

• At the school, teachers from an urban background are unaware of the great value of local technology and knowledge. They are teaching the children that progress consists of replacing traditional practices by modern ones.

Figure 1 Model of the Andean cosmovision

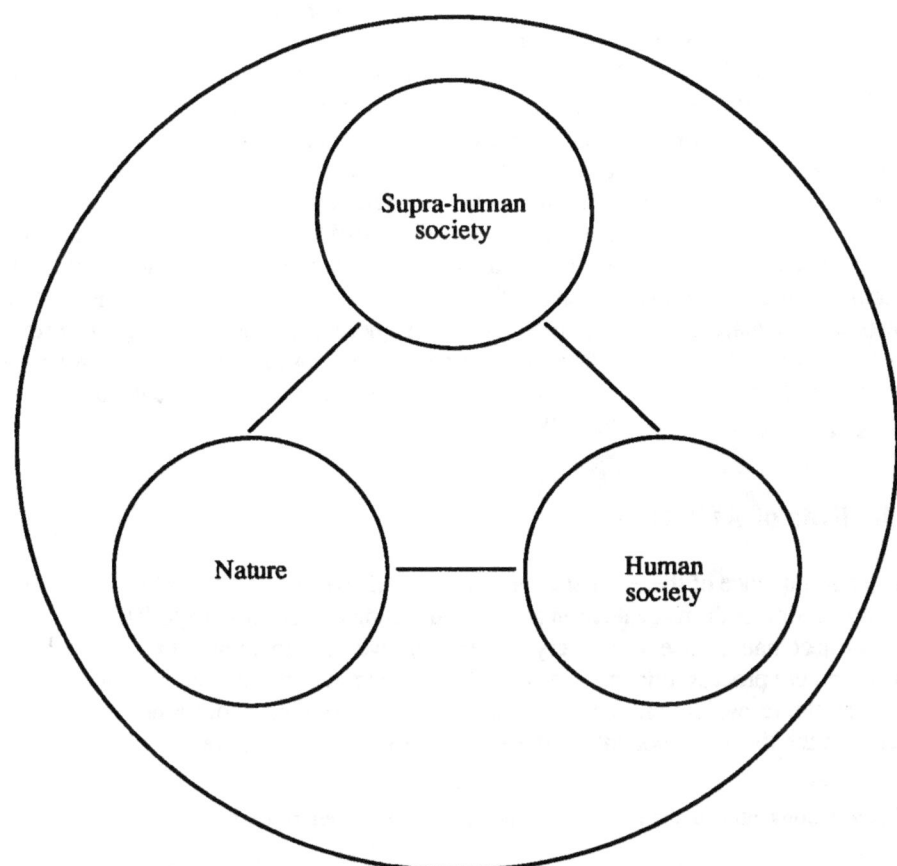

Source: van den Berg (1990)

- Many religous sects seek to discredit local traditions and social systems for mutual help, without considering the real needs of the people.
- Rural extension institutions, despite the high social motivation of many of their staff, are promoting the use of Green Revolution technology, considering local technologies and knowledge as one of the main reasons for lack of progress.
- Many local people are migrating to Chapare, a major coca-producing area. This is a strategy to diversify income. This zone is being transformed into an area where the 'Western way of life' prevails. In the effort to create economic alternatives to the coca bush the government is promoting high-input agriculture, which is taught to the

immigrants. On return to their communities, many of them think that they can also use this approach to increase production in their traditional lands.
• Many farmers seek to get more out of their farms in order to satisfy new needs, such as a radio, a television, a bicycle, and so on.
• Following the drought of 1983, Bolivia began receiving considerable quantities of food aid from the developed countries, consisting mainly of wheat and sugar. This has led to a gradual change in nutritional habits, undermining the demand for local foods and leading to the decline of locally adapted technologies.

To sum up so far, we have learnt that the technologies used in farming systems have to be understood in the light of a dynamic equilibrium of the three elements of the Andean cosmovision—human society (community), nature (ecosystem) and the supra-human world. Indigenous technologies tend to disappear because modern development approaches are based on a materialistic 'cosmovision' which ignores the transcendent, giving priority to technical and economic factors over social and spiritual ones. The erosion of indigenous knowledge can be halted only by re-evaluating the Andean cosmovision and bringing its three components back into balance.

The Role of AGRUCO

As a consequence of these experiences, the activities of AGRUCO had to be redefined to allow room for the re-evaluation and revival of indigenous knowledge. This also meant the project had to see itself very differently: we are no longer the centre of the development process; this role is assumed by the communities themselves. We can only support the re-evaluation process, with the aim of bringing about a new equilibrium between the three components of the Andean cosmovision (Figure 2).

Figure 2 Conventional and new understanding of development process

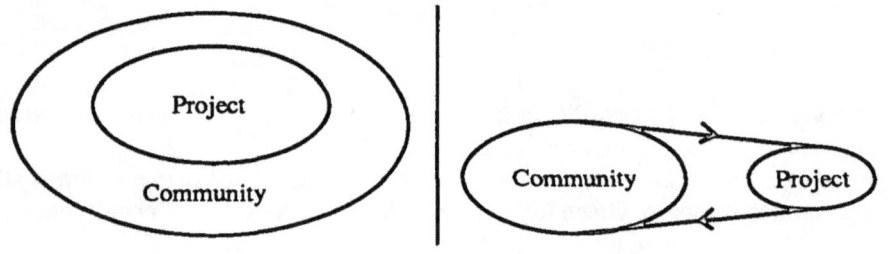

A. Conventional project-centred development approach

B. External project supporting the strengthening of indigenous knowledge and creativity

Accordingly, we have divided and organized our activities into four interlocking areas: academic education, participatory research, support to rural communities in different agro-ecological zones, and support to rural development institutions (government and non-government) (Figure 3).

Figure 3 Activities of AGRUCO

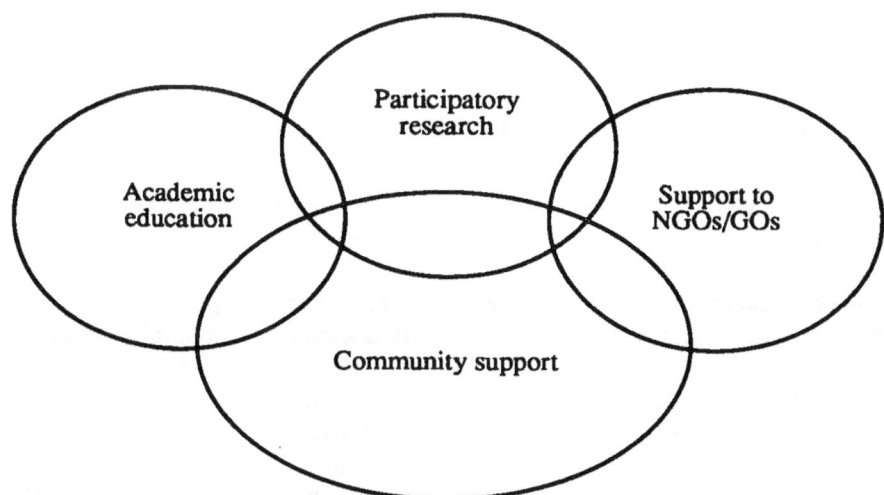

AGRUCO tries to establish a holistic approach, ensuring feedback between the four areas of activities. To provide the world of academic education with the necessary materials, it must carry out activities in support of the community's self-defined development agenda, which provides the context in which its own professionals can learn about the relationship between community, nature and the transcendental. At the same time the results of this participatory research are important for the education of professionals at the university, since once working for development institutions they will be able to initiate and apply this approach themselves.

Another important consequence of this approach is the fact that the communities' knowledge is directly and indirectly integrated into that of the university. Academic education is normally very theoretical, far removed from the realities and requirements of rural communities. AGRUCO has transferred three of the six curricular subjects to the communities. As part of these three subjects the students have to live for a week in the community, where they are taught and trained in participatory research, methods of community support, and planning and monitoring methods. Included are participatory rural appraisal techniques directed not only towards diagnosing problems but also towards identifying components of the traditional knowledge and technology system of the communities. In essence, the work of AGRUCO resembles that of a network, linking the various actors in rural development such as universities, development institutions and rural communities, which normally function in isolation from one another. The network tries to establish horizontal relationships of equality between outside agents and rural communities, taking into account that two different cosmovisions are interacting and that neither of them has the right to subordinate the other. This interaction creates the basis for rural development.

Network Activities at Community Level

In this section we will summarize some of the most important methodological instruments which AGRUCO is using and developing at community level.

Before we start we should note, however, that the networking concept is one we apply only in our interactions with other outside agencies (universities, government and non-government organizations, etc). We do not apply it within the community because we perceive the community's existing structure to be rather more complex than that of a simple network. This is also why we do not create any parallel organizational structures, such as specific sub-networks or a committee of farmers interested in indigenous knowledge. We respect the community's own organizational structures, seeking only to strengthen these where appropriate. Despite the absence of any externally created network, formal or informal, some basic principles of the networking idea can nevertheless be applied, as we will describe in the following sections.

Handbills

Indigenous knowledge and the practices, technologies, social relations and rituals associated with it are increasingly marginalized, found only in regions less affected by modernization. The *campesinos* frequently travel to urban centres, where they are exposed to modern knowledge promoted by the media, schools, church, and so on. Even in more isolated communities, where there is still a rich tradition of indigenous knowledge, we find considerable differences in knowledge between the older and younger generations.

All this makes it important to document and redisseminate these disappearing elements of indigenous knowledge. AGRUCO therefore began to apply and adapt a method developed by a Peruvian non-governmental organization (NGO), consisting of the production of handbills on traditional technologies (Rist, 1992). The process of producing the handbills is highly participatory and firmly based on the *campesino's* perception of needs. Box 1 provides further details.

> **Box 1. Preserving indigenous knowledge through handbills**
>
> Due to climatic changes in the past few years (droughts followed by floods, unusual early frosts) many *campesinos* believe that the traditional weather forecasting indicators are not valid any longer. As a result crops are now sown according to the availability of labour rather than the expected beginning of the rainy season. This can cause very low yields. Other important factors accounting for low crop yields are the introduction of chemical fertilizers, pesticides and so-called improved seeds (mainly of potato), which has encouraged the farmers to abandon the use of organic manure and to intensify and modify the crop rotation, reducing fallow periods and moving away from mixed cropping. The result has been a drastic reduction in soil fertility.
>
> Against this background the professionals of AGRUCO began designing a series of handbills on the management of soil fertility, mixed cropping, and weather forecasting indicators (botanical, zoological, atmospheric, astronomic and ritual), based on the knowledge of the few *campesinos* who still knew these technologies. The handbills are very straightforward and use the words and expressions of the *campesino* when describing the technology. This information is complemented by some sketches.
>
> Handbills are presented by a professional and a *campesino* together in draft form at the monthly meeting of all community members or, if this is not possible, at some other event, such as a training course. The presentation generates a discussion among the participants, confirming or changing the information on the draft. At the same time the advantages and disadvantages of the practice are discussed. This approach helps the *campesinos* to realize the importance and value of their own knowledge.
>
> Often the presentation of a handbill leads to significant corrections, after which the final version, validated by the community, can be produced.
>
> The next stage is to disseminate the handbills. The final version is reproduced and as many copies as are required by the *campesinos*, the community authorities, teachers, doctors and others are supplied to the originating community. The final version is also presented at a meeting of AGRUCO staff, ensuring that the 'new' elements of indigenous knowledge are known to all and available for use in daily work. Finally, the handbill is introduced to the Andean Network on Indigenous Knowledge (PRATEC), which has its headquarters in Peru and a decentralized research unit in Bolivia, managed by AGRUCO.

Networking between communities

Since nearly all the activities of AGRUCO are carried out within the rural communities, relevant elements of indigenous knowledge can easily be detected. The quantity and quality of indigenous knowledge differs considerably from one community to another. This makes the exchange of knowledge between communities highly relevant.

We therefore seek to bring together *campesinos* who have a certain problem with others who do not, because they are still using their traditional knowledge. This facilitates a reciprocal, horizontal learning and dissemination process. Box 2 provides an example.

> **Box 2. Learning from each other: Technology exchange between communities**
>
> In 1989/90, in one of the communities where AGRUCO was working, many potato plots were severely affected by a disease called *verruga* (*Synchytrium endobioticum*). The farmers said that the disease was due to the 'tiredness' of the most affected seed varieties, because for year after year they had used the same varieties without ever refreshing them, as far back as they could remember. They therefore asked whether AGRUCO could help them obtain new, fresh seed. AGRUCO agreed to this, but not without warning them that the seeds sold at the local market were usually a mixture of different types, which might not lead to an improvement in potato seed quality. The peasants were well aware of this but did not see any other solution, since they could not remember from where their fathers had brought the now 'tired' potatoes. Knowing that the easiest way is not always the best, we insisted that we would be willing to help obtain fresh potato seeds, but only if they were willing to investigate, in exchange, the origin of their own potatoes. This was eventually accepted.
>
> Some time later the farmers informed us that their potatoes had come from a community called Chapisirca, quite some distance away and at an altitude of about 4300 metres, some 300 metres higher than their own community. Chapisirca's higher altitude and relative isolation should reduce the level of plant diseases and pests, with the result that excellent seed quality could be expected. Farmers and AGRUCO staff visited Chapisirca immediately. Some of the members of the community at Chapisirca remembered hearing that, many years ago, *campesinos* of their community used to take potato seeds to their relatives in Japo, and that these were needed to refresh their germplasm. They were willing to sell some of their excellent potatoes, thus not only providing the peasants of Japo with refreshed seed but at the same time re-establishing a social relationship that had considerable technical, environmental and economic significance. Through the inter-community contact an important in situ germplasm conservation method had been rediscovered which, from now on, will permit the refreshment of potato seed whenever this is needed, independently of outside help.

Networking between community and outsiders

In accordance with its principal objective, AGRUCO has been assisting development institutions interested in modifying their work through the exploration of indigenous knowledge.

In many cases we found other institutions willing to change their approach because they had already felt disatisfied with the gap between their aims and the true interests and needs of farmers. Sometimes, institutional change could be effectively supported by bringing together farmers who have been working with AGRUCO over the past few years with farmers who were still distrustful.

The effectiveness of this horizontal diffusion of indigenous knowledge is illustrated by the experience described in Box 3.

Box 3. Gaining confidence through horizontal contacts

In 1990/91 the Gesellschaft für Technische Zusammenarbeit (GTZ) began a joint project with the government's Regional Development Corporation to improve food security and livelihoods in the province of Arque. This area is strongly affected by soil erosion, causing a continuous decrease in productivity which has driven many farmers off the land.

The only way to arrest and reverse this trend is to apply techniques to prevent soil erosion, such as the construction of terraces and other physical and hydrological changes to the landscape. The corporation started a large-scale information and training campaign in the communities, with the aim of motivating the farmers to initiate soil conservation activities. However, in spite of all their efforts and the incentives offered, no work was done on soil conservation for about 2 years. Then AGRUCO suggested that a delegation from Arque should visit the neighbouring province of Tapacari, where some communities had been working with AGRUCO on soil conservation over the past few years. The *campesinos* agreed to this and soon afterwards two delegations were organized and travelled.

As soon as they arrived they observed and discussed the experiences of the local *campesinos*, who had already evaluated various soil conservation techniques. The visitors agreed with their hosts at Tapacari that something had to be done about soil erosion, confessing that their reason for not doing anything was their fear that once their soil had improved the project would take away their land to repay the funds invested in helping them to improve it! This in spite of the fact that the project's professionals, aware of these rumours, which often spread when a new project starts up, had repeatedly explained to the *campesinos* that this could not and would not occur.

The farmers of Tapacari told the visitors that they had had the same fear at the beginning but that, seeing the advantages of soil conservation, they had decided to go ahead. They had felt that, should this problem ever arise, they would fight with all the more energy and determination since they would be defending not only their land but also all the hard work and sweat they had put into its improvement.

Two weeks later the first families who had participated in the visit began organizing to make terraces, without even asking for the project's help. After that other groups expressed a wish to make a similar visit, including a group of women.

Participatory research at community level

Practically all AGRUCO's research is carried out in the same communities as its support activities, so the research activities are part of the support and vice versa. Furthermore, the participatory research is also part of the academic education, since AGRUCO not only teaches a 'pre-specialization' in agro-ecology and sustainable development at the university but also grants a fellowship each year to some 8 to 10 students of different disciplines who, after concluding their studies, conduct research in the field as part of their graduate studies. The students stay with AGRUCO for a year, doing participatory research in one of the communities where AGRUCO works, under the guidance of one or two professionals of the AGRUCO team. During their stay the students also learn about project administration, studying participatory planning, monitoring, evaluation, logistics and financial management.

These research activities are of two kinds, applied and basic. Applied research is directed towards solving one of the problems experienced by the community or evaluating some of the indigenous technologies scientifically. This latter work has a twofold function, first to provide evidence of the scientific basis of traditional technologies, and secondly to improve and disseminate these technologies at community level through the implementation, monitoring and evaluation of the experiment, together with one or more farmers, on their own plots. Once concluded, the research work must be given back to the community in the form of a training course, engendering discussion and analysis among the farmers. The results are then available for use in other communities if appropriate.

Basic research is directed towards a better understanding of how the Andean cosmovision influences social organization and the use of knowledge and technologies in the community. This research is very important for a project that seeks to support the community in its evolution.

Through this participatory research we have found that the *campesinos* are very interested and spend a lot of time thinking and talking, not merely about their own lives but about where they stand in relation to nature and the rest of the world. (This is a characteristic that can also be found in all their rituals, in which the supra-humans are asked to re-establish the equilibrium between society and nature, to stop or prevent wars and shortages of food, not only for them but for the whole world.) Their reflections on the current situation help them re-establish their relationship with nature, since they compare the present with the past, which helps them to find new answers for the future.

Academic education in and with the communities

As already mentioned, three of the six subjects taught by AGRUCO at the university are taught in one of the communities, which means that the students must spend 1 week living with the families of these communities. This gives students the opportunity to study the realities of rural life and to learn not only with their intellects but also with their hearts, by feeling and seeing.

The fact that the *campesinos* are the teachers of their own reality had a surprisingly positive effect on the strengthening of their own identity. This was expressed as follows by one of them, during a speech at the closure of the course:

> Thank you for bringing so many students to our community, so that they can see and learn how we live and what we need and thank you all who have listened, when we told you what we know, because at this moment we feel very proud of our own knowledge and we are giving it freely to your hearts so that you shall always remember us.

Networking with Outside Agencies

The aim of AGRUCO's relationships with other agencies is to show them how an intercultural dialogue can serve as the basis for a joint development process.

With other universities
AGRUCO is a small part of just one of the nine universities of Bolivia. It has signed an agreement with five of the other universities, all in the Andean areas. These agreements are facilitating the participation of a limited number of students in the three subjects taught in the communities. AGRUCO also sponsors and organizes every year an intensive training course for university teachers.

With government and non-government organizations
Professionals in many projects are now becoming aware of the need to change away from the conventional development approach based on the Green Revolution. As a result there is increasing demand for support in getting out of this development trap.

AGRUCO therefore saw a need for a new area of activities directed towards both government and non-government organizations. The former are less important these days due to the drastic reduction of their human and financial resources in the context of structural adjustment. Private-sector institutions are therefore all the more important, being often the only form of development assistance present in many Bolivian communities.

A small sub-network was created, whose 25 members—mainly private development institutions—are interested in integrating the aims of AGRUCO into their work. Each member has a specific task in accordance with local needs and capabilities. A special three-phase methodology has been developed to enable NGOs to diagnose needs in the rural community and respond to them.

First a delegation of AGRUCO staff visits the area where the NGO to be assisted is active. Together with all the staff of the NGO, AGRUCO staff maintain continuous contact with the *campesino* families over a period of 5 to 7 days. In this way they get to know both the NGO staff and the farm families. At the same time they study the way in which the NGO interacts with the community, and the results it has achieved.

Impressions are summarized daily in a joint session with the staff of the NGO being visited. NGO staff serve to set the critical self-evaluation process in motion. At the end of the first phase, again together with the entire staff, there is a joint planning session in which future activities are written down.

After 3 to 4 months the second phase begins with the joint evaluation of the activities carried out following the planning session at the end of the first phase. Successes and failures are evaluated together with the farm families, and new activities are again planned. During this phase participatory planning and evaluation techniques can be practised.

Students interviewing a farmer

The third and last phase is timed to coincide with the NGO's normal planning schedule. AGRUCO's support consists of helping staff integrate the experiences of the previous two phases into the planning of all future activities. This presents a further opportunity to practise and strengthen participatory planning.

Andean network for handbills on indigenous knowledge

As we already mentioned, once a handbill on indigenous knowledge has been finalized, it is sent to the International Andean Network, where handbills from various countries are collected for further editing, processing and distribution. The main objective is to facilitate the exchange of indigenous knowledge among all participants. Through this collection, a great deal of little known indigenous technology is made available, helping to solve problems where local knowledge has not found a solution. Over 600 handbills have now been processed. A list of these, or of course the complete collection, can be purchased from AGRUCO.

References

Berg, H. van den. 1990. La tierra no da asi nomas. HISBOL–UCB/ISET. La Paz, Bolivia.

Rist, S. 1992. Desarrollo y participación: Experiencias con la revalorización del conocimiento campesino en Bolivia. Serie Tecnica No. 27. Proyecto AGRUCO, ICUMSS. Cochabamba, Bolivia.

Address

Stephan Rist, AGRUCO, Casilla 3392, Cochabamba, Bolivia.

Networking with Women Farmers

Janice Jiggins

Introduction

Women farmers are thought to produce more than half the world's food crops and a large proportion of its non-food crops and livestock as well. But it is unfortunate that even participatory approaches to agricultural research and development through networking do not ensure women's participation. Powerful methods exist for optimizing women's participation. Yet methods alone do not suffice.

If women are to participate actively in shaping the agenda, managing the process and sharing the costs and benefits of research and development, something different has to happen. There must be an explicit commitment to work on issues that women consider important. A core group of female staff is vital. Care has to be taken that the methods used in working with communities are sensitive to women's ways of communicating and styles of leadership; the timing of meetings, interviews and other tasks should take into account farm women's work burden and working hours, which are not the same as those of a 'nine-to-five' researcher. Including women changes the agenda set by men, modifies social relationships and attitudes, and influences the outcome of research and development.

It is a basic principle of changes in voluntary behaviour that the people involved 'own' both the problem and the solution. It is a basic principle of building sustainable organizations that participants see their own interests sufficiently reflected in the organization's mandate and activities, and are rewarded in proportion to the effort they put in. If farmer networks ignore or undervalue women's contributions, neither the technical nor the organizational process will prove durable.

This paper discusses examples of the following situations: spontaneous informal networking arrangements among women farmers; externally assisted (by government or non-government organizations) networking with women farmers, mostly by men; and formally constituted autonomous women's networks, run by and for women.

Spontaneous Informal Networking

Women farmers' informal networks have not been studied much. Yet, to focus on just one aspect, they are often the primary vehicle for maintaining the genetic viability of farming operations. The bulk of the seed and planting material used by women farmers is replenished and augmented through women's informal seed exchange networks. In many parts of the world this exchange is supported by women's roles in trading in local markets. The networks can take many forms, as the following examples illustrate.

Seed exchange on the Kore Rice Scheme in Kenya

In the early 1980s, most women rice farmers holding tenancies in the Kore I Rice Scheme in Kenya used mixed seed retained from the previous harvest, but some used pure, improved seed they obtained by working for (male) relatives who had plots on the nearby Ahero Pilot Scheme, where pure seed was more freely available (Povel-Speelers, 1982). Over time, woman-to-woman sharing of the pure seed has become the main channel for the spread of pure seed among women tenants on the Kore I Scheme itself. Three somewhat interdependent types of organization are involved: women's kinship networks; women's informal work groups; and women's personal friendships in their neighbourhoods. None of these interact with formal seed distribution channels.

Maintaining the viability of soya bean seed in East Java

Farmers in East Java find it difficult to maintain the viability of soya bean seed in storage. To ensure a flow of fresh planting seed, farmers in different villages informally arrange to exchange soya beans between fields planted at different seasons of the year, with farmers in particular villages specializing in dryland cultivation during the rainy season (Soegito and Siemonsma, 1985). The exchange is typically mediated by women, who are responsible for post-harvest seed management.

Exchanging common bean landraces in Malawi

Women bean farmers in northern Malawi maintain a large portfolio of bean varietal mixtures, including named varieties indigenous to specific communities as well as more widely distributed types (Martin and Wayne Adams, 1987). They maintain these sets of landrace mixtures even though pure lines are available at affordable prices. The set of beans in a mix varies in a broad north-south cline, probably due to both environmental factors and consumer preferences.

The mixes are composed both by random mixing of the physical components and by deliberate selection on the farm. Gifts and market exchange are also used. Colour, shape, size, cooking qualities and taste, susceptibility to pests and diseases, plant architecture and length of growing period are among the characteristics used by women farmers to distinguish their seeds. A number of the types are unstable, freely out-crossing. It is not known how the women maintain, over the generations, a set of varieties which remains palatable and continues to meet their criteria.

Externally Assisted Networking

The following examples highlight some of the key managerial, social and technical aspects of externally assisted networking with women farmers.

Seed dissemination in Zimbabwe

The Zimbabwe Seeds Action Network (ZSAN) focuses on community-level production and storage of seed of traditional maize and small grains in the drought-prone areas of Zimbabwe. Farmers' groups—the majority of them women's groups—have been cooperating for more than 5 years in the testing and evaluation of indigenous grain varieties, and in the management of warehouses. The income accruing from sales of the seed from the warehouses is retained by the managing group. The network is supported by ENDA-Zimbabwe (an affiliate of the international, South-based non-government organization, Environmental Development Alternatives), and local Zimbabwean non-government organizations (NGOs).

Although the majority of the farmers in the drought-prone areas are women, the majority of the staff of ENDA-Z, including those most directly involved in supporting the ZSAN, are men. This fact highlights what appear to be two contradictory lessons: (a) there are men who are committed to, and skilled at, developing effective networks among rural women; (b) although women benefit from such networks, network activities in themselves do little to challenge or influence the gender relations of power which contribute to women's poverty. Unless women's leadership is deliberately encouraged, women's place within networks tends to be a derived or subordinate position, as in ZSAN. In turn, the presence of an effective core group of female staff seems to be important in getting empowerment adopted as an organizational policy goal.

Farmers' groups in Botswana

The Agricultural Technology Improvement Project (ATIP) in Botswana began in the late 1980s to experiment with what came to be called farmer research groups (FRGs) and farmer extension groups (FEGs) as a way of institutionalizing collaboration between researchers and farmers in research design, field trials, evaluation and extension activities (Norman et al, 1988).

Women-only and mixed membership groups were tried, as well as working with existing groups and specially formed groups. No type of group proved more suitable than others; the ones that failed did so for reasons other than these criteria. Today, the FRGs and FEGs have evolved into routine partners of the national farming systems programme. They allow a far greater range of technology options to be tested (many suggested by the farmers themselves), create opportunities for farmer involvement in the formal technology development process, and provide a forum for direct interaction among farmers, researchers and extension agents. As the programme has matured, farmer-to-farmer field days, farm walks and demonstrations are taking place spontaneously, as well as more formally with the support of the agencies concerned.

Over half the membership consists of women farmers. Women have proved particularly good at sharing the results of research through the FEGs, because of their multiple informal links in their area (principally via the family, schools, clinics, other women's groups, and trading).

Lessons from Kenya

A pioneering study was carried out of women's groups working with agricultural extension advisors in Kenya (Muzaale with Leonard, 1982; 1985). Although all groups referred to have now been registered by the Ministry of Social Affairs, some of them emerged spontaneously as a grass roots initiative. Others were promoted by government agencies or NGOs. The extension services now designate women's groups as important contact points. Important lessons can be drawn from these experiences for working with and supporting networking activities among women's groups:

- The members were somewhat more literate than women in the surrounding populations. Further, groups whose leaders were better educated had a stronger capacity for self-management, becoming quite independent of the supporting agencies by the end of the project.
- The majority of members were not living permanently together with a husband or grown sons. This increased both their need to seek support through group activities and their freedom to participate.
- Most groups started with non-agricultural programmes (whether self-organized or stimulated by a non-agricultural agency). Contrary to what has been found elsewhere, no problems associated with a group's particular programmatic history were observed.
- The key criterion for continuing enthusiasm was the members' capacity to convert external support into concrete benefits.
- Most of the groups were welfare-oriented in their objectives, seeking to meet wide-ranging family needs. The ability to earn income, and not just increase production, was thus important.
- Poorer members dropped out as environmental stress increased through a prolonged period of drought. At the end of the drought period, the groups had lost over half their membership and the surviving members all rated themselves as more famine-resilient.
- The survivors did not expect the others to return quickly to the group, reckoning that the participation costs (principally time, labour and money) were too high.
- Further analysis revealed a strongly seasonal dimension to the costs of participation. Women operating with fewer resources are particularly exposed to seasonal stress (the labour required of them at harvesting, for example).
- The analysis also revealed that communal networking intensified the inequitable distribution of participation costs. More networking among individual members and individual farmer-to-farmer visits, for example, might have introduced greater flexibility.

Formally Constituted Women's Networks

Formally constituted women's networks are a relatively new feature of the agricultural research and development scene. Initiated and run by and for women, they have sprung up in all parts of the world, at local, national, regional and even global level. They are typically constituted as movements, with little hierarchical control, a great deal of local autonomy and a flexible leadership as women shift in and out of organizational roles in accordance with family and marital obligations. What is significant is that, through these approaches, they are able to address biophysical and socio-economic concerns at the lowest point of human organization and at the same time influence policy at higher levels.

The WorldWIDE Network

The WorldWIDE Network was founded in 1981. Its concern is environmental management and protection by and for women worldwide. Women who share the goals and values of WorldWIDE are encouraged to form local chapters or fora, to explore common themes and concerns and to define and execute locally appropriate action. WorldWIDE's local success stories include, for example, the following from Manyu, Cameroon:

Food output had declined due to soil erosion as former conservation practices had fallen into disuse. Many young men had left to seek urban jobs and there was no longer the labour available to look after the land in the old ways. A local woman organized women's groups through the church and the market-place to practise soil conservation. They met together on agreed days to cut and plant contour bunds on each other's farms. They shared techniques of composting kitchen waste and animal manure. The network of groups has expanded throughout the district and is now seeking other ways to promote sustainable farming practices and to share seed and other planting material.

In November 1991, over 400 WorldWIDE members met at the Global Assembly of Women and the Environment in Miami, to share their experiences and formulate recommendations to the United Nations Conference on the Environment and Development (UNCED) in Rio de Janeiro. Their recommendations were incorporated the following week in the deliberations of the global meeting of over 1200 women, who represented women's movements and networks in both the North and South. The meeting's concluding statement was submitted formally to Maurice Strong, Director-General of UNCED, as an input to Agenda 21.

Special Characteristics of Women's Networking

One of the characteristics which is helping women's networks to emerge as fora with a distinctive voice within the development debate is the incorporation of explicitly feminist ethical and intellectual values. These include respect for the diversity of all life forms, an appreciation of the holistic way in which biophysical and socio-economic phenomena are related in time and space, and an attitude of caring for others rather than seeking to control them.

In terms of principle, feminist positions on many issues are close to those of many ecologists. They can be clearly distinguished from the reductionist, relatively short-term and profit-driven practices which continue to dominate technology development in agriculture and related fields. This is not to argue that these practices are wholly and invariably inappropriate or ineffective, but rather that there is a need for a more balanced approach, derived from a wider process of dialogue and shared learning, and combining the best elements in each school of thought. The characteristics of the 'feminist-ecologist' position can be and indeed sometimes are strongly upheld by men, but they are, perhaps, more apparent in women's networks, being rooted in women's distinctive experiences of life, biologically and socially. For most rural women there is little difference between the things they do to feed the family and the things they do to care for it. The organization of daily life is a seamless web of tasks designed to maintain the system as a whole.

Women optimize and exploit biological diversity to provide a continuous flow of food to the household, typically growing or using a portfolio of plant and other species, including those not even recognized by agricultural professionals as crops. While modern agriculture dismisses or eliminates these plants as weeds or inferior land races, women's experience warns them that the extreme uniformity displayed by modern agriculture is in fact dangerously vulnerable to sudden disaster. Once the years of infancy are past women, in their own lives, are more vulnerable than men to biological and social changes of state and status that transform their relationships with the surrounding world.

They use their influence in the family and community (engaging in what is often pejoratively dismissed as mere gossiping) to create and maintain the informal relationships without which the performance of common agricultural tasks and the maintenance of natural resources would be impossible. They also recognize that there can be no sustainable farming in the long term without sustained communities, which means in turn that safeguarding the regenerative capacity of both the natural environment and the human species is important. Scientists such as Vandana Shiva view the regenerative principle as quintessentially feminine, the origin of evolutionary diversity and creativity. This is in contrast with the hybrid products of commercial agriculture, which produce sterile seeds for uniform environments.

Another characteristic of women's movements is their insistence on networking as the means to secure 'power-to-act' rather than 'power-over', as the case of women in the Australia-wide LandCare movement illustrates. Although the LandCare movement is not a women's movement as such, women's ability to motivate people seems to be a key factor in persuading people to make the individual effort to change their practices and attitudes, and to join in community-based activities to transform agriculture's interaction with the environment (Roberts, 1987; Mills, 1991). Women's skill in gaining the cooperative support of technical experts 'has been central to the success of many LandCare groups. In many cases the female touch had led to a networking between interested parties, without the competitive or exclusive leadership style which too often has plagued rural organisations in the past' (Roberts, n.d.).

For a further discussion of the characteristics of women's networks, I refer the reader to 'feminist advocacy' and the 218 success stories that are documented in Martin-Brown (1992).

Conclusions

Supporting farm women's networking activities requires sensitivity to a number of aspects of women's lives, priorities and modes of interaction. Although the details vary from place to place, these might be summarized as:
- Women's need to secure benefits which do more than providing material gains. Often, the potential benefits sketched in project proposals are unobtainable in practice, for the requirements further constrain rather than increase the flexibility women seek in order to deal with seasonal stress, family crises and the multiple daily requirements of survival.
- Women's preferred styles of communication, interaction and leadership. These emphasize the interactive and the collaborative, and the free mixing of social knowledge and technical knowledge in information exchange and discussion. They also respond to the constraints of women's longer working days, home-based responsibilities, and intermittent free time.
- The difficulties of incorporating women's priorities and capacities into existing male-dominated networks.
- The challenge of accepting that power relations and the agricultural development agenda will change as women find their voices and build their organizational strength through women-only networks.

References

Martin, G.B. and Wayne Adams, M. 1987. Landraces of *Phaseolus vulgaris* (Fabaceae) in Northern Malawi: I. Regional variation; II. Generation and maintenance of variability. *Economic Botany* 41 (2): 190-203, 204-215.

Martin-Brown, J. and Ofosu-Amoah, W. (eds) 1992. Partners in Life: Proceedings of the Global Assembly on Women and the Environment, 4-8 November 1991, vols 1 and 2, UNEP, Washington D.C.

Mills, D. 1991. Women in LandCare: Nurturing an ethic. In: Runs on the Board: Proceedings of the Annual Conference of the Soil and Water Conservation Association of Australia.Toowoomba, Australia.

Muzaale, P. with Leonard, D. 1982. Women's groups and extension in Kenya: Their impact on food production and malnutrition in Baringo, Busia and Taita-Taveta. Report to the Ministry of Agriculture. University of California, Berkeley and Nairobi.

Muzaale, P. with Leonard, D. 1985. Women's groups and extension in Kenya. *Agricultural Administration* 19 (1): 13-28.

Norman, D., Baker, D., Heinrich, G. and Norman, F. 1988. Technology development and farmer groups: Experiences from Botswana. *Experimental Agriculture* 24 (3): 321-331.

Povel-Speelers, F. 1982. The incorporation of women farmers into the organisational structure of the Kore Rice Schemes. Discussion Paper, Irrigation and Drainage Branch, Nairobi, Land Resources Development Division, Ministry of Agriculture.

Roberts, B. 1987. *Mother Earth: The role of women in developing land stewardship in Australia.* Queensland, C.W.A. Oakey.

Roberts, B. n.d. Women in rural conservation and LandCare. University of South Queensland. Toowoomba, Australia.

Soegito and J.S. Siemonsma 1985. On-farm programme Wonorejo. Soybean Improvement Paper I, ATA 272, Malang, Indonesia, Quarterly Technical Progress Reports 13, MARIF.

Address
Janice Jiggins, De Dellen 4, 6673 MD Andelst, Netherlands.

Farmer Networking at International Level

Frances Kinnon

Introduction

Probably one of the oldest structured farmers' networks, and the only body representing national farmers' organizations at international level, the International Federation of Agricultural Producers (IFAP) was founded in 1946. Since then it has grown to include 81 farmers' organizations and cooperatives from 57 countries, over half of which are from the world's developing regions.

Initially preoccupied with the reconstruction of agriculture after the devastation of the Second World War and guided by a strong belief in the philosophy of active international cooperation, IFAP has gradually shifted its emphasis as the farming sector has undergone unforetold changes, particularly in terms of productivity. Industrialized countries are currently facing a state of structural oversupply, severe competition and increased trade tensions on world markets. The emergence of developing country economies and their associated agricultural problems has broadened IFAP's perspective and spheres of activity, especially with their accession into membership of the Federation.

One bond which has remained constant for almost all members is their attachment to the family farm unit. These units vary from intensively managed small-scale farms in Japan to large intensive ranches in Australia and Argentina, and from near-subsistence farms in Africa and Asia to highly capitalized commercial farms in Europe and North America.

Approach and Operations

Why an international farmers' network?

IFAP was established to secure international cooperation between national organizations of agricultural producers in meeting the nutritional and consumptive requirements of the peoples of the world and in improving the economic and social status of all who live by and on the land. Although strongly conditioned by the situation at that time, this central statement in the Federation's Constitution is as valid today as when it was first drafted in 1946. The fact that conflicts are bound to arise among competing producers the world over in no way invalidates this first mandate of IFAP. Despite their traditional independence and individualism, farmers—as a numerous yet often physically isolated group—need to speak with one voice at certain times, even at an international level. Hence the continuing need for a forum of exchange and a consensus building platform such as IFAP.

IFAP aims to accomplish this broad objective in four ways:
° By acting as a forum at which the leaders of national farmers' organizations from different countries can meet to develop an understanding of world farming problems, to discover mutual interests, to make recommendations and to take coordinated action.
• By promoting and strengthening independent representative organizations of agricultural producers from grass roots to national level throughout the world.
• By representing the world's farmers, bringing their concerns to the attention of international meetings of governments and other bodies which have a bearing on farmers' interests and livelihoods.
° By keeping members informed about international events and issues of concern to them.

Structure of the network
IFAP can be called a worldwide farmers' network in the fullest sense, since it is entirely governed and financed by its member organizations, which are themselves the national farmers' organizations in their respective countries. IFAP is democratically governed by its members, each of whom is on an equal footing so that no single organization or group of countries or set of special interests can gain a dominant position. This is guaranteed by the Constitution, which requires that, in administering the Federation, a suitable balance is maintained between the different interests represented within it. The independence of IFAP is further guaranteed by the fact that its mainstream activities are exclusively financed by the annual subscriptions of members, who not only have a say in how the Federation is run but are also in a position to resist any outside attempt to influence its policy. IFAP has no political or religious allegiance and accepts into membership any nationally representative independent farmers' organization.

Such organizations, which include farmers' unions, farmers' cooperatives and chambers of agriculture, must be autonomous, self-financing and self-governed, with those holding office freely elected by members. A more recent consideration is that the organization should be at least 2 years old—an indicator of its sustainability. Applications for membership go before the Executive Committee. If supported by two members, the organization is accepted on a provisional basis pending confirmation by the Constitution and Membership Committee, which meets at General Conferences.

Criteria for membership allow for representation by more than one national organization from a country. Thus, for example, Brazil, the United States, France, Italy and Cameroon are represented by several organizations. There is also provision for membership by national committees. This facilitates membership for developing countries in which farmers' organizations are not strongly developed. Individual organizations may be unable to pay the minimum annual subscription, but by forming a national committee they can share the cost among them. One vote per country is permitted in official procedures such as elections.

The vast majority of the farmers linked together through their national organizations within IFAP are self-employed and essentially small- to medium-sized. IFAP now represents virtually all agricultural producers in industrialized countries and several hundred million farmers in developing countries. Inevitably, in the first years of the Federation those organizations that joined tended to be long-established—some of them pre-dating this century, many of them export-oriented. In recent years, however, and especially during the 1980s, there has been a marked increase in membership from the developing countries, most of which represent smallholder subsistence-oriented farmers.

Forum of exchange: The World Farmers' Congress

The governing body of IFAP is its General Conference, which is held every 2 years. The Conference gathers together representatives of all its member organizations as well as observers from international and intergovernmental agencies. These biennial meetings are working, policy-making conferences. They comprise a maximum of 300 delegates made up of the elected and decision-making members of national organizations. All delegations are led by active farmers with direct knowledge of, and concern for, the problems affecting their members at national level. Thus discussion at the international level is still farmer to farmer.

Nearly all developing country farmers' organizations face financial problems which prevent them from participating in IFAP events using their own resources. For this reason, through the support of development cooperation agencies such as the Canadian International Development Agency (CIDA), the Norwegian Development Agency (NORAD) and the Swedish International Development Agency (SIDA), a series of seminars for developing countries has been held just prior to the World Farmers' Congress since the early 1980s. These events have allowed the in-depth examination of an issue of specific concern to developing countries as well as ensuring the participation of developing country farm leaders in the major plenary session debates at the General Conference itself.

The World Farmers' Congress fulfils several funtions. It examines the major problems facing agriculture worldwide and jointly seeks solutions; it establishes ties of friendship and solidarity among farmers' organizations around the world; it identifies the priority actions of the Federation over the next 2 years; it deals with constitutional and membership issues and elects officers for the next 2-year mandate.

Sessions traditionally cover the world farming situation and trade issues. Other themes are generally of a broad focus, including subjects such as the development of rural areas and industrial uses of agricultural products.

Inevitably, serving such a broad membership can give rise to difficulties, especially with regard to controversial issues. Perhaps the most difficult challenge recently has been achieving a meaningful consensus on international trade issues, particularly the Uruguay round of the GATT negotiations. The views of IFAP members differ depending on whether their countries are net food exporters or importers. It is on potentially divisive issues such as this that IFAP has served as a forum for increasing understanding and appreciation of the views of others. By arranging regular consultation with the GATT Secretariat and negotiators, IFAP has involved farmers in the discussion process.

Other issues have undeniably served to unite members. The catastrophic fall in world prices of tropical commodities, the impact of structural adjustment on the farming sector, and the debt burden are issues with which all developing country farmers' organizations can strongly identify. At a global level, all farmers' organizations share concern about the rising pressure on farm incomes. The percentage of the food dollar that goes to farmers continues to decline, with the increasing concentration of power both upstream (input supplies) and downstream (processing industries, etc) of the farmer imposing fixed prices. In addition, environmental issues have become a priority for practically all members in recent years.

Promoting and strengthening farmers' organizations

With the growing importance of developing countries in the membership of IFAP, it was recognized that farmers in these countries were operating under a very different set of policies than their counterparts in industrialized countries. Usually working within severe social and economic constraints, farmers' organizations—where they existed at all—often lacked adequate administrative and organizational structures. Often physically isolated and politically marginalized at home, such organizations were almost entirely lacking in opportunities to communicate and interact with each other at regional and international level.

Bearing in mind IFAP's specific mission to encourage the formation and support of independent farmers' organizations, the Federation launched a development programme specifically targeted to the needs of such organizations in developing countries. This institution- and capacity-building exercise has focused on three objectives: strengthening farmers' organizations where these already exist; encouraging their formation where they do not yet exist; and ensuring their participation in the international debate both at IFAP events and at other international fora.

Strengthening existing farmers' organizations has involved activities to increase their representativeness and enlarge their support base. Especially important has been encouraging the participation of women farmers and small-scale producers, who constitute the majority of farmers in developing countries. Efforts have also been made to enhance their representational capacity, through improved information and communication. Fostering financial independence and self-support through income-generating activities and, in certain cases, promoting independence from government have also received attention. 'Towards self-supporting farmers' organizations' was the theme both of the 1990 developing country seminar and of a recent publication.

Where national level organizations do not yet exist, their formation has been encouraged principally through increased cooperation and coordination between existing regional or local farmers' groups. It has been amply demonstrated that the structuring of farmers' organizations from grass roots to national level can be a long and meticulous task. Attempts to speed up the process often lead only to superficial groupings without solid foundation, which disintegrate once external support is withdrawn.

Ensuring the participation of farmers' organizations in specific IFAP and other international events has been the third focus. IFAP has held many workshops and seminars on issues proposed by members, such as pricing policies, marketing and input supplies, and the structuring of farmers' organizations. Regional IFAP meetings promote greater interaction between farmers' organizations sharing similar backgrounds and facing similar problems. These have aimed to identify policy positions on specific issues. While providing a unique opportunity for the exchange of ideas and experiences, they are also decision-making fora. Positions emerging from such sessions are regularly conveyed to appropriate international bodies as well as disseminated through IFAP publications.

Linking farmers with policy and research

IFAP has also tried to promote consultation and collaboration between the intergovernmental agencies and farmers' organizations. For example, in collaboration with the International Fund for Agricultural Development (IFAD), three regional workshops were held in Kenya, Tunisia and Nepal on the themes of farmer participation in development projects and the strengthening of farmers' organizations. IFAP has also organized three international consultations: one with the International Institute of Tropical Agriculture (IITA) in Ibadan, Nigeria, another with eight international research bodies based in Nairobi, Kenya, and a third consisting of working sessions with nine national agricultural research centres and the United Nations Environmental Programme (UNEP).

Representing the world's farmers at international level

As the only worldwide association of farmers, IFAP is recognized as official spokesperson for the farm sector at many international fora. The Federation is entitled to send representatives to UN meetings, to participate in the debates as an observer and to circulate written statements in response to UN documents. It also enjoys other privileges, such as consulting with the UN Secretariat, conducting special studies of mutual interest, and so on.

IFAP has recently participated in several relevant intergovernmental conferences. For example, its representatives attended the FAO-den Bosch Conference on Sustainable Agriculture and Rural Development and the International Conference on Nutrition, both held in 1992, and the recent Conference on the Economic Advancement of Rural Women sponsored by IFAD. IFAP's President addressed the United Nations Conference on Environment and Development (UNCED) held in Rio de Janeiro in 1992.

IFAP acts as the spokesperson for farmers at big international meetings

Networking as a federation: Greater regional focus
With regard to networking among members, an important trend in recent years has been the emergence of a stronger regional focus. This movement started in the early 1960s, when the European representatives decided to set up a committee on which problems specific to European agriculture and policy might be discussed. This initiative was followed in 1985 by the formation of an African Regional Committee. Since its

establishment this committee has successfully served as a launching pad for improving contacts between African farmers' organizations at regional level and for concerted action with regard to representation to intergovernmental bodies and institutions. This year there will be a further research consultation in Manila, the Philippines. The objective of these consultations is encourage agricultural research programmes to address the needs of small farmers and women farmers.

Two other regional initiatives concerned strengthening links with farmers' organizations in South-East Asia, launched in 1989, and a regional conference for the Farmers of the Americas, held in 1993. The latter brought together farmers' organizations from North, Central and South America to discuss the impact of regional trade agreements on the farming sector.

Through IFAP's general and regional activities, the leaders of farmers' organizations get to know each other, exchange views, discover common interests and develop contacts. This generates a considerable amount of activity which is carried out member-to-member, without the intervention of IFAP. For example, exchange meetings are set up between different national farmers' organizations, or presidents are invited to address each other's conferences. More recently, North-South exchanges and collaboration have become more common, especially regarding advice on the structure and internal management of national farmers' organizations. Examples include bilateral exchanges between farmers' organizations in the Netherlands and Burkina Faso and the Philippines, between Sweden and Kenya and Zimbabwe, and between the United States and Nicaragua.

Towards sustainable agriculture

Environmental issues are not new to IFAP. For example, in 1974, IFAP's Group on the Environment selected four areas deserving priority attention. These were the positive environmental role of agriculture and forestry; land use and management; the control of chemicals; and information and education. Owing to limited staff resources in the headquarters Secretariat, the Group decided to address this problem by networking through IFAP's member organizations. They recommended that one member organization act as a focal point for the exchange of information, experience and news, and that such exchange be conducted entirely by mail, owing to the impossibility of convening meetings between conferences. Periodic or occasional digests of the data collated were to be circulated to all IFAP members.

For about 10 years, trade and other issues took the limelight away from environmental questions in IFAP. Then, in 1986, the IFAP General Conference discussed as a major theme the conservation of soil and water. The report of the Policy Committee states:

> Farmers in many countries recognize the problems involved in preserving soil and water resources but often lack sufficient information on how to deal with them effectively. There is a need for constructive technical and economic advice and guidance. Farmers' organizations have an important role to play in promoting research, in establishing

information networks and advisory and extension services and in training farmers to make more efficient use of their resources. These requirements are also valid regarding the use of fertilizers and pesticides.

The subject of pesticides, especially their safe use and handling by small farmers who may have little or no formal education, has been a recurrent issue of concern to leaders from developing countries, and it had been mentioned in the developing country seminar held just prior to the same General Conference. Special emphasis was placed on the need for more research on local forms of pest control, especially biological methods. This theme was taken up 2 years later at the next developing country seminar, which focused on the development of more appropriate farming systems.

During the mid- to late 1980s farmers' organizations began to find themselves working in a new social context. Not only was government policy shifting towards less state intervention. People's movements, especially those associated with consumer, development and environmental issues, were having a growing influence on the public debate. Of particular concern to farmers in both the industrial and the developing countries was the fact that debate was often taking place without their participation, so that their concerns and interests were not being taken into consideration. Farmers realized that if they did not take the initiative on environmental issues they would no longer be setting their own agenda but instead would have policies thrust upon them. No longer the only actors in rural areas, nor the only group concerned about environmental issues, farmers' organizations realized that they would have to form coalitions with these other groups.

Preparations for UNCED acted as a further catalyst, focusing the attention of farmers' organizations on the broad range of issues that come under the heading of sustainable agriculture. With UNCED in mind, and given the need to establish farmers as key players in the debate about sustainable agriculture, IFAP held a major conference on agricultural development and the environment in October 1991, in Iceland. Using contributions from some 50 national farmers' organizations throughout the world, IFAP drew up a policy statement which outlined the roles and responsibilities of both farmers and society in achieving a more environmentally sound agriculture. The statement made the following points:

° Only an economically viable agriculture will be capable of achieving sustainability and other environmental objectives.

° Farmers need to be rewarded for their role in managing the countryside.

° A shift towards the use of environmentally sound technology, as well as education and training in the use of this technology, is essential.

° A reasonable time period, as well as financial incentives to farmers, are important in adjusting to new rules concerning the protection of the environment.

° Consultation with and involvement of farmers through their own representative organizations is critical to achieving sustainable agricultural production. This in turn requires a favourable climate for the autonomous development of representative farmers' organizations from grass roots to national level.

The latter point deserves particular stress. IFAP's member organizations have always insisted that the adoption of environmentally sound practices cannot be imposed from above. The harnessing of the individual efforts of millions of agricultural producers throughout the world is crucial. This is best done through a network of representative farmers' organizations on a worldwide basis.

Environment committee

As a follow-up to UNCED, on the initiative of its Swedish member organization IFAP established an Environment Committee. The work of this committee is structured round three international meetings to be held in Europe, America and Asia, each dealing with a specific theme. What distinguishes these meetings from other IFAP conferences is that, in addition to inputs from members, they will draw on advice from specialists in intergovernmental organizations, universities and international environmental and development agencies. This is to provide the committee with as broad a perspective as possible from which to draw its conclusions and on which to base its recommendations.

The outcome of this series of hearings will be a consensus document. This document will aim to relaunch agriculture by placing the farmer at the centre of a policy for sustainable economic, social and ecological development. Although 'sustainable agriculture' will form the core of the document, many overlapping and adjacent issues will also be covered. These will include rural-urban balance, the distribution of responsibility for achieving more sustainable agriculture among different social groups, the question of renewable resources, the use of agrochemicals, biotechnology, natural resource management, trade and the environment, and research and education.

In addition to setting the framework for the future work of the committee, the first of the sessions, held in Stockholm, focused on the theme: 'The message from Rio: Demands and expectations on farmers and their organizations'. Invited to address the meeting were high-level representatives from the Food and Agriculture Organization (FAO), which had opened the discussions on sustainable agriculture and rural development issues at the Den Bosch Conference of 1992, and the United Nations Development Programme (UNDP), which has a major part to play in implementing Agenda 21, especially with regard to capacity- and institution-building.

The second session, held in Washington in April 1993, covered issues which will have an impact on the future of farming in developing and industrialized countries alike. Both population and biotechnology, as they relate to sustainable agriculture, were discussed in detail, as were the implications of a 'greener' GATT for trade and the environment. For the first time, the Consumers' Union was invited to address a formal IFAP meeting. In presenting its views on 'safe food', the union stressed the importance to farmers of maintaining consumer confidence in this area.

The last of the sessions, to be held in the Philippines in October 1993, will bring together leaders of farmers' organizations from Africa, Asia and Latin America. This session will have a developing country focus, dealing with the pressing issues of poverty alleviation and the environment, especially with regard to erosion, desertification and the problems of fragile lands. The suitability of Green Revolution technologies will be re-assessed.

Conclusions

Like other sectors of the economy, farming is experiencing a severe depression, hit especially by macro-economic factors such as high interest rates, inflation, and depressed world prices for commodities. Against this background there is a special need for farmers' organizations to be pro-active, to set their own agendas and to give active support to environmental programmes. The short-term thinking of politicians can be devastating for a sector such as agriculture, in which the care of natural resources—its production base—requires a long-term perspective.

IFAP feels that the main obstacles to solving environmental problems are not necessarily in the technical field but rather in the political, economic and social domain. If this is true, farmers' organizations have a vital role to play in promoting sustainable agriculture by lobbying politicians and educating the public. This is one reason why IFAP has consistently stressed the need to establish farmers' organizations in those countries where they do not exist, and to strengthen them where they do. However, a major institution and capacity-building exercise such as this will require an active partnership between farmers and intergovernmental organizations and a firm commitment to the farming sector on the part of national governments.

Much to the frustration of governments, extension workers and the voluntary organizations, farmers in industrialized and developing countries alike listen to their own organizations more than to any other source of information. Working with and through farmers' organizations is therefore one of the best ways of disseminating new ideas or information. However, no matter how well informed and educated a farming population is, it will take time for ideas to be adopted and farming practices modified. For their part, farmers' organizations will need to examine the question of networking not only with other farmers' organizations but also in a broader context. Building up 'rural partnerships' with consumers, environmental groups and others active in the rural areas, in a balanced, open and constructive relationship may be one of the best paths to sustainable agricultural development.

Farmers' organizations must also play a different kind of role to that of the past, taking the lead and making demands on society, on politicians and on scientists to provide new technologies, systems and resources. Farmers' networks are not cast in stone: to survive and flourish, they must be flexible enough to adapt to a rapidly changing external environment. Failure to do so will lead to declining enthusiasm for the network. For this reason, maintaining regular contact with members is essential.

Some farmers among the grass roots membership are not yet convinced of the importance of environmental issues. Others are confused by conflicting messages coming from the media and from other interest groups. For these reasons, farmers' organizations must offer leadership and a clear strategy in this important area. Cooperation and collaboration among farmers' organizations at the international level will help to reinforce these efforts.

Address
Frances Kinnon, Secretary for Rural Development, International Federation of Agricultural Producers, 21 Rue Chaptal, 75009 Paris, France.

PART THREE

NGO Networks

Daring to Share: Networking among Non-government Organizations

Paul G.H. Engel

Introduction

In recent decades non-government organizations (NGOs) have invested a great deal of time and money in networking. Numerous formal and informal networks are the result. At the same time, the study of networking as a social phenomenon has received increasing attention. Social scientists have noted how people 'capitalize on' their social relationships in order to deal with the challenges of life. Following theorists like Bourdieux (1991), some would say we have finally understood the importance of investing in 'social capital'.

In my view the new emphasis on networking has brought us numerous advantages, but some disadvantages too. The advantages stem from the fact that networking entails explicit recognition of ourselves as social beings. Our knowledge, technologies and practices are not created by individuals in splendid isolation, but socially, as a result of interaction with each other. Our fascination with networking for development purposes reflects the fact that we no longer feel that there is only a single source of knowledge for dealing with a given problem, but rather that there may be as many sources of knowledge as there are people involved—and people of different walks of life too: scientists are just one of many communities trying to come to grips with development issues.

Finally, our investment in networking is intimately connected with our concern for sustainability. As is painfully clear from the various theatres of war in the world today, sustainable development can only be achieved where people have worked out a way of living with each other. In other words, it can only be built only upon sustained social relationships. We can no longer point at either farmers, or policy makers, or researchers, or development workers, or even money lenders, as the prime culprits for what is wrong with agriculture. To be successful, our analysis and actions will have to involve all of these groups. Current efforts at networking are at the forefront in helping effective relationships to emerge.

The disadvantages have to do with the fact that we haven't yet been able to conceive of networking in such a way that it can permanently and successfully compete for resources with our other important activities. In other words, most NGOs and, I might add, most donors still see networking as an overhead, to be kept to the bare minimum. In my opinion such a point of view is at odds with achieving sustainable, socially integrated development.

This is the argument I would like to advance in this paper. In doing so I cannot and do not pretend to be exhaustive. I can only draw on the networking experiences with which I am familiar. I hope my perspective will contribute to a better understanding of networking, and to the improved management and evaluation of networks.

First I will describe what I mean when I say 'networking' amongst NGOs, and why I think networking is work. In the process I will try to say why I believe networking deserves a place of its own in our budgets—not as part of administrative costs, training, field work or documentation and information, but as a separate line item, a complementary set of activities which creates a specific added value to whatever else we do. Second, I wish to draw out some of the central issues in the current networking experiences of NGOs as documented in this book. Third, I will try to define the added value of networking, and make some suggestions for developing a conceptual framework for evaluating network performance. This, I hope, will help us make the networking process more manageable.

What's It All About?

My primary interest is not with networks but with network*ing*—the process resulting from our conscious efforts to build relationships with each other to further the cause of sustainable development. Networks are the more or less formal, more or less durable relational patterns that emerge as a result of such efforts.

One of the most intriguing questions about networking is: what exactly does it contribute to our work? And how do we recognize this contribution when it occurs? One may safely assume that networking makes a great contribution to NGO work the world over. The proliferation of networking activities is proof enough of that. Yet, what exactly is this contribution? Why is it so difficult to put our finger on it? Padrón (1991), one of the outstanding networkers of Latin America, was among the first to recognize the need for more systematic analysis, precisely because it is so difficult to establish what networking is, why it happens, and how its advantages can best be used to develop the NGO community's efficiency.

From his own vast experience with both thematic and institutional networks in Latin America, Padrón suggests a central thesis for understanding NGO networking: networking is about *sharing*. And he warns: sharing may be one of the most demanding requirements in development work, yet it is the most essential common denominator developed by the poor in order to provide for each other and live under adverse conditions. 'Daring to share', as he puts it, 'is neither easy nor automatic. It requires a willingness to be open-minded, it requires having enough confidence in one's own work to expose it to others, and at the same time, the necessary humility to understand one's position as one among many.'

In my view, this makes networking more than simply working together—more than the mere collaboration of individuals and institutions on the basis of common interests. Networking has to do with achieving 'social synergy', as Haverkort and Ducommun (1990) put it. Networks represent 'communities of ideas', a space for like-minded people to interact on the basis not only of common interests but of conflicting ones too, building mutual trust and learning to accommodate each other's needs. The core businesses of networks are not so much the manufacture of products and/or the provision of services, but social learning, communication, and the making of meaning. In focusing on 'mind' rather than 'matter', networking adds a fundamentally new quality to human cooperation. It enhances inclusive thinking, creativity and dialogue.

As a consequence, networking activities show fundamentally different characteristics to more product- or service-oriented ones. A good example of such a characteristic is redundancy in communication between people. In product- or service-oriented organizations, this is checked and, if possible, eliminated. Meetings, informal discussions, coffee breaks, personal mail and instructions are to be kept lean so as to avoid stealing time from productive work. Networking, in contrast, is an activity in which we positively indulge in dialogue, are encouraged to exchange ideas and experiences, are urged to take the time to listen to each other and to work towards a new way of understanding old problems. On the face of it unproductive, this actually provides a space for reflection, for breaking down barriers and stimulating creativity. As we will see, this can often lead to a considerable increase in the quality, if not also the quantity, of our work. A certain amount of redundant communication may well be a prerequisite for bringing about such improvements (Engel, 1990).

As far as I'm concerned, any attempt to manage networks which overlooks this fundamental characteristic is doomed, for it misinterprets the reasons for networking, the social needs and forces which lie behind it. This is not to deny the importance of specifying products and services in the realm of networking. Indeed, as we will see, this is important too, since it provides indicators for evaluating networks and measuring their success. But the understanding of networks can never be reduced to the simple 'production' logic so commonplace in institutional thinking today. The added value of networking is strongly tied to the development of ideas, to shared experiential learning and to making sense of the world through communication. The challenge is to develop a framework for planning and evaluating networking activities that is concrete enough to be serviceable but that does not lose sight of this fundamental issue.

The Central Issues

Several key issues emerge from the experiences of NGO networking documented in this book. In this section I will refer to these experiences, and also to those of the global

Non-formal communication is essential

networks El Taller and Huridoc. (El Taller is a Tunis-based foundation with a membership of 100 NGOs worldwide; Huridoc is a long established international human rights organization.)

What all these networks have in common is that they have reached a stage of consolidation. All of them have survived the uncertainties of institutional infancy and matured into respected adolescence, carving out a niche for themselves in the local, regional and/or global NGO community.

My main interest will be in looking for the value added to NGO activities by networking. In the view of the NGOs themselves, what makes networking worthwhile? I will look into this issue by raising three questions:
- What triggers networking amongst NGOs?
- What makes networks take a more permanent form?
- What activities characterize networks?

What triggers networking among NGOs?

NGO networks appear to develop when NGOs themselves, or members of their staff, perceive a lack of access to relevant knowledge to be a critical factor hampering their work. This lack is not looked upon as absolute. On the contrary, it is perceived as being surmountable if a sharing of ideas, experiences and information is organized among relevant parties.

In Tamil Nadu and Pondicherry States of India, NGOs and farmers agreed that many sound traditional practices existed which needed to be brought to light and were worth disseminating (see Quintal and Gandhimathi, p.177). In the case of the Andean Council of Ecological Management (CAME) in Peru, severe droughts and floods convinced NGOs that they were unable to respond adequately to the needs of Andean farmers. They attributed this failure partly to their lack of familiarity with Andean and other appropriate technologies, partly to the inadequacy of current ways of managing climatic risks and partly to the absense of inter-institutional coordination (see Manrique et al, p.167). Participants at an Oxfam workshop in Cotonou identified the isolation of local project staff as a handicap to their work, giving rise to the formation of the Arid Lands Information network (ALIN) (see Graham, p.271).

In some instances a more general lack of coordination appears to have stimulated networking efforts. This was the case with the Association of Church Development Projects (ACDEP) in northern Ghana, which arose from the perception that church projects operated in isolation and tended to replicate each other, offering similar and sometimes competing services within the same locality. In this case coordination was based on the vertical administrative structure of the church. Coordinators could not provide the necessary technical back-up, and different approaches to development existed without the benefit of a learning process between projects. Under these circumstances networks are required to assume a broader role, facilitating organizational integration and change. This may lead to the establishment of new specialized units or agencies dedicated to specific tasks in support of all the NGOs concerned.

Similar perceptions account for networking initiatives at the global level. The networking activities of Huridoc were based, according to its evaluators, on the premise that information about human rights is hard to obtain, difficult to dissiminate and essential to the protection of human rights (Tajaroensuk et al, 1992).

However, if a lack of knowledge or coordination alone is identified as the motive for networking, this falls short of recognizing the full urgency of the intentions behind NGO networking efforts. In an increasing number of cases these efforts spring from the growing awareness of NGOs that their need is an acute one—far more than just a casual search for knowledge. Networking in such cases is a response to the wish to put heads together, to join forces, to search jointly for new ways of understanding and intervening in circumstances that are complex and defy simple analysis. Often, it is triggered by a rejection of mainstream or conventional thinking, by the struggle to articulate an alternative, more sustainable approach to development. In the case of CAME, as we have just seen, the NGO community realized that its grasp of Andean and other appropriate technologies was inadequate. In India, the environmental problems caused by modern technologies led to the search for alternatives that would be more sustainable. At a global level, the founding members of El Taller experienced a similar sense of having been let down by conventional approaches. In the words of its Secretary-General, Sjef Theunis: 'El Taller was born from the need for reflection voiced by NGO leaders from around the world. Women and men who work at the heart of their society are feeling that citizens and politicians have lost their direction and focus' (El Taller, 1990). In some cases a theory or slogan is adopted to provide guidance in developing an alternative approach. Whether this is called low-external-input and sustainable agriculture, or environmentally sound agriculture, or ecological agriculture, the effect is the same: a banner is raised and serves to rally the troops.

A third factor that often triggers networking among NGOs is the wish to participate in the public and/or government debate about development and so to influence policy making. One reason for the creation of CAME was the wish of its members to transcend their isolation and make themselves heard at regional and national level. NGOs have realized that matters of policy are beyond the scope and competence of any single one of them, and that they must combine forces to achieve an impact.

With respect to the first two factors triggering networking efforts, most of the networks in the cases presented in this book show a remarkable degree of similarity. With respect to their involvement in the policy debate, however, there is more diversity. CAME and the Red de Agricultura Ecológica (RAE), in Peru, aim explicitly to contribute to the debate. Other networks, such as ALIN in Africa and those in Tamil Nadu and Pondicherry States of India, seem to put much less emphasis on this, at least for the time being.

To sum up the answers to my first question, networking efforts appear to be triggered when three perceptions are widely shared by NGO leaders, staff and clients:
- A lack of access to the knowledge of others is hampering effective performance and causing specific problems.
- At a deeper level, there is a need to gain a more comprehensive and more subtle understanding of the complex problems NGOs are dealing with, and to create new ways of supporting grass roots development.
- The experiences of NGOs at grass roots, and the interests of the poor on whose behalf they work, need to be voiced at national (or higher) level, in order to contribute to the formulation of more effective development policies.

The first perception leads to the wish to *upgrade* the performance of NGOs through collective action. It leads networkers to emphasize the sharing of ideas and experiences, whether through meetings, communications technology or documents. The second impression leads to a wish to move *upstream* in terms of both analysis and activities. In so doing NGOs question the very relevance or efficacy of field operations themselves, 'reaching beyond the evident consequences of the problem at hand to address its source', as Korten puts it (p.25). This process emphasizes shared diagnosis, reflection, the making of sense and meaning, and coordination at a strategic level. The main concern is to achieve a better paradigm of development—a challenge seen as beyond the powers of any single agency acting alone. This need emerges especially clearly now that NGOs carry major responsibility for developing more sustainable alternatives to the conventional chemical-based technologies that have dominated agricultural development in recent decades. Accordingly, the third impression leads to what may be termed an *upshift* among NGOs. In shifting the focus of their activities, NGOs give expression to the need to articulate alternatives and lobby for them through the media and in the corridors of power.

All three U's reflect, in one way or another, the desire to improve the quality of NGO work and the contribution NGOs make to rural development. However, each network reflects a specific blend of these ingredients. From local networks of service-oriented NGOs, interested mainly in the practicalities of upgrading their performance, to global strategic networks, concerned almost entirely with advocacy and directing their efforts to a specific cause, no two networks combine them in the same way or to the same degree.

What makes networks last?
Why do some networking activities lead to the establishment of institutionalized networks, while others do not? Many networks have been designed and initiated but have quickly petered out as the initial momentum was lost and (prospective) members reverted to business as usual. Yet many survive, and it is these 'sustainable' networks that can teach us lessons about the conditions under which networking activities become more institutionalized and less casual.

Before going into this, we have to deal with the often heard argument that networks function best when they are informal and for that reason no attempt should ever be made to institutionalize them. This is exactly why we have to distinguish between networks and networking. Every individual, every organization engages in building relationships with others, that is to say in networking. Most of these relationships remain informal—personal and rather subject to chance. Some, however, acquire such relevance to the life and work of the individual and the institution that these decide to formlize them in order to guarantee them a more permanent future and a 'place' in the institution's life, including, perhaps, offices and other facilities. Arguing that networks should always remain informal is akin to saying people should eat, but never build a kitchen.

Formal networks, then, are neither a prerequisite to, nor the necessary outcome of, networking activities. Under what conditions do networking relationships become more permanent and take the form of an institutionalized network?

For the NGO networks described in this part of our book, a first condition is that many people must share the view that networking will add value to their work. These people, moreover, must be in a position to articulate such views and to develop a mission for the network. This means that the organizational mechanisms for formulating a shared mission must be in place. That is, the question of who may or may not be a constituent actor, having the right to co-determine the ground rules or constitution of the network, has to be answered, and a procedure agreed upon for developing a shared perspective, or a 'theory of poverty' as Tim Brodhead puts it (quoted in Korten's paper in this book, see p.25). Often such questions are not dealt with very explicitly by those constituting a network. And as most networks start off very informally, they do not have to be. Networks tend to evolve around a closely knit group of charismatic leaders. Initially at least, they determine who is 'in' and who is 'out', and set the agenda for network activities. However, when networks become larger, the need to develop more transparent and more broadly participatory ways of taking such decisions arises.

This brings us to a common point of origin for all formal networks. They start with a phase of planned activism on the part of a 'motivator group', as Manrique et al call it—a phase in which, first, ideas are exchanged, then a few trial activities lead to recognition of the value of sharing with others, then one or a small group of enthusiastic 'prime movers' promotes the idea of networking and, finally, a meeting with prospective network members is organized. During this phase, a lot is done, but often in a rather unplanned fashion. The outcome is usually a workshop or a meeting at which, among other things, the idea of forming a network is discussed and agreed on.

The extent to which this phase can be spontaneous and unsystematic depends to a great degree on the scale of the operation. While national and even regional NGOs may organize for a network in a very informal way, for international efforts such as El Taller it tends to take years of programmed activities to prepare the foundations for the network. Yet, though the scale differs, the mechanisms seem pretty much the same: the combined efforts of a group of prime movers, network facilitators and prospective members lead to the formulation of ideas, plans and activities which eventually result in the establishment of the network. Prime movers are the people, generally leading members of respected NGOs, who create the idea, the vision on which the network is to be built. Network facilitators are those who, by virtue of the space and time allowed to them by their own organizations, engage in actual networking—organizing and supporting a first set of activities closely attuned to the needs and wishes of the prospective members of the network. In some cases prime movers and network facilitators are the same people. Mostly, however, facilitation is the function of an embryonic secretariat attached to one of the prime movers. The planned activism phase always requires a direct or indirect sponsor to cover at least some of the operational costs.

During this phase a number of issues arise. First, the importance of communication and participatory methods is directly felt by the participants. If the network is to embrace a wide group of NGOs and their staff, these must participate intensively in the formulation of its objectives and the organization of its initial set of activities. Often, this is easier said than done. For those working in isolated rural areas, especially, taking the time to share ideas with others working elsewhere and developing the habit of doing so is, however enriching, far from axiomatic. Networking must compete with other activities on an already overcrowded agenda. And there may be severe communication problems, even with the next-door village, let alone across national or regional boundaries.

A still more difficult but essential task is the development of a shared conceptual framework which facilitates the exchange of ideas and experiences. Kolmans, in describing experiences in Peru, notes the unrealistic goal setting and the extensive theoretical discussions that took place during the first year of preparations (p.151). But he also indicates why these were necessary: to overcome ignorance on the topic of environmentally sound agriculture among prospective members; to 'work through' one-sided viewpoints, such as the idea that all that is traditional and Andean must be sustainable; to integrate a social science perspective into the technology generation process; and, last but not least, to convey to donors and other potential supporters the actual needs of rural people. In my view, what Kolmans is referring to is a classic case of 'making sense' of the idea of setting up a network—checking the real need for it, defining its potential to support its members, adapting the basic concept as more people provide ideas and inputs. This process takes a lot of time, yet it is an essential preparation for the network to be. It transforms a set of diverse people and organizations, each with a different opinion and an ill-defined sense of common purpose, into a like-minded group with many interlocking relationships and a shared perspective, thereby enabling them to start learning from each other.

This process of acquiring a common understanding and a shared purpose or mission is in all cases linked closely to the existing activities of prospective members in their respective areas. The immediate needs arising from the field work of each of the founding institutions provide the network's basis and its raison d'être. From the very beginning, the network is intended to support the work of the NGOs involved. Only if this support, or the potential for it, is directly perceived by members can they assess the added value of networking and set this against their other obligations. And only then can the principle of reciprocity apply. Or in other words, as Manrique puts it in describing the CAME experience, if an NGO fails to contribute to the network, 'there is no networking, no network'.

The phase of planned activism is possibly the most difficult phase for a donor to support. Because no shared frame of reference, values and discourse have yet been developed, the network, in so far as it exists at all, will not be able to articulate its processes, services and products in a satisfactory way for an outside audience. What is needed during this phase is sponsorship—the provision of seed money from an institution that is prepared to be a prime mover without interfering too much in the detail of plans and preparations. Rather than 'knowing' the network is going to be a success, the sponsor should be open-minded, believing or hoping that this will be the case. The provision of seed money by ILEIA to support the nascent Tamil Nadu network is a good example of sponsoring.

An old Dutch proverb seems to fit network building nicely: ' a good beginning is half the job'. To succeed, networks must have firm foundations. These are best laid not by pushing things along as fast as possible but by taking one step at a time.

To sum up, from the cases reviewed here, the following factors appear critical to the successful establishment of formal networks:

- Planned activism, facilitating and supporting (*never* replacing or ignoring) the existing activities of members.

° The will and the opportunity to discuss, negotiate and agree on the mission of the network in a way that is transparent and agreeable to all or most of the prospective members.
• A cast of actors, including prime movers, network facilitators, prospective members and sponsors, willing and able to carry the networking process through its initial, ill-defined phase.
• Broad participation of prospective members in the design and implementation of initial activities.

What activities characterize networks?
Networks span an enormous range of activities: from field trips to communication by satellite or electronic mail, from project planning to education and training, from editing a newsletter to organizing a conference, from lobbying ministers to admonishing a member for the late delivery of data, to name but a few. This is one of the reasons why it is hard to define networking as a phenomenon. Since the activities of many networks are discussed and illustrated in detail in the different contributions to this part of the book, I will not repeat them here. Instead, I would like to try to categorize them.

I suggest that networks generally concentrate their efforts in four clusters of activities: the provision of services; learning together; advocacy; and management.

The provision of services refers mostly to information and training. In providing or commissioning services, the network seeks to make optimum use of the capabilities and facilities of its members, supplementing these with inputs from elsewhere when necessary. A needs assessment and/or a diagnosis of strengths and weaknesses among network members often serves as the starting point. Typically, the network secretariat is attached to the member organization considered most capable of running its most important services. The service function is supported by what might be called the network communications infrastructure. Almost all networks have a newsletter, which acts as a major vehicle for the exchange of ideas and experiences. Documentation and library services are often provided as well, as also is the development of training materials. Publications are not limited to the proceedings of events, nor to technical matters and development issues. Methodological and project support documents are often a high priority as well.

Services may expand into other domains, including technical consultancies (as in CAME), product certification (as in RAE), or the coordination of input supplies (as in ACDEP). The common denominator in the services provided by networks is their responsiveness to members' needs. In addition to the general emphasis on training and information, network-specific packages of services may therefore evolve.

Learning together embraces all the joint activities undertaken to raise members' level of understanding of the complexity of development problems. These may include mutual appraisals, exchange visits, workshops and other meetings. Sometimes permanent working groups on specific topics are formed. The emphasis varies from network to network, but common elements are diagnosis, exchange, comparison and synthesis. Many networks stress, as ALIN does, the importance of visits and workshops, not as ends in themselves but as the starting point for reflection. Diagnosis and the making of an inventory of available technological and methodological options are often part of the process. CAME and RAE aim for gradual synthesis, and the standardization of scientific and technical approaches may also be involved (although this is more typical of research networks).

Advocacy refers to those activities performed or facilitated by the network on behalf of its members and clients that enable them to participate in the public or government debate about development policy. It is the defence of the interests of members and clients, particularly when these are, or were previously, disadvantaged members of society. Advocacy involves the network in formulating proposals on contemporary development issues and voicing these to government and/or in the public media. The network may also organize conferences on controversial issues, contribute articles to scientific journals, or distribute relevant publications to key decision-makers. Coalition building with relevant parties from outside the network, or with other networks, is often on the agenda as well. The advocacy function of NGO networks is not currently as widespread or as transparent as their learning and services functions. For example, NGO leaders chose not to include advocacy among the tasks of the El Taller network, feeling that political lobbying was something that had to be done by individual members, on a case by case basis.

However, as Korten (p.25) points out, the theory and practice of strategic networking is making considerable progress among NGOs at present. And there is indeed a potential for conflicts of interest between a service orientation and the advocacy function. Yet this need not be a matter of 'either/or'. Advocacy and services are two sides of the same coin. How could we possibly do without either of them? What does seem to happen in the more permanent networks is a greater emphasis on the services and learning functions, particularly during the early stages of network development. The dedication to advocacy in a network is very much a matter of the personal choice and initiative of its leaders.

Finally, the management function consists of facilitating the networking process. This includes maintaining or improving its communication infrastructure, overseeing its operating procedures, monitoring its resources, activities and outputs, and linking with other organizations and networks. Without going deeply into this function, many aspects of which are discussed in detail in other papers, let me point out some common characteristics of network management today.

First and foremost it is important to emphasize what is *not* being done under the management function. Networking secretariats are kept lean, delegating as many tasks as possible to member organizations. The decentralization of functions and the autonomy of members is emphasized continuously. A directory of members and their organizations is often among the first fruits of a new network. It is generally motivated by the wish to facilitate networking without having to go through the secretariat. The network facilitators' mandate usually stems from a meeting of prospective members who decide to initiate a more formal networking process, but it is generally a mandate to advise and support, not to organize and command. Most networks decide not to engage in the management of funds for members, however expedient this may seem at a certain moment. In the words of Manrique (p.167), this can turn the network into a 'battlefield for acquiring money'.

It seems important to define clearly the composition, responsibilities and prerogatives of the network board, secretariat and, if applicable, implementing bodies. The degree to which the secretariat or hub of the network should engage in implementing activities itself is an issue that frequently arises. Whether formal rules should replace the largely unwritten rules that govern operations during the early stages also tends to be an issue. Although it is difficult to generalize, experience suggests that a degree of formality is desirable in larger, older networks, and that the secretariat should have a mandate to take decisions on membership, on the provision of advisory and other services, and on monitoring and evaluation issues, particularly where these are sensitive.

Even if network activities are mostly delegated to members, they still require time and money. The moment networks become more permanent, therefore, the issue of fund raising comes up. During the early days, prime movers free up the energy and other resources required for networking from somewhere else, often from within their own programmes. Donors enter the scene only when the contours of the new network have already been delineated. This means that during the early stages exchange and communication is often limited to those who are able to provide the necessary facilities and funds themselves. This limits the participatory process precisely at the stage when broad participation appears most desirable, even mandatory. Sponsorship during the early stages can thus make an important contribution to ensuring broad initial participation.

How do these four categories of network activity help us in understanding the networking process? From the above analysis I conclude it is particularly the emphasis on learning together that sets successful NGO networks apart. Networks, as it were, are 'learning organizations' by definition. They are designed and operated to break down isolation and facilitate social learning processes among actors within the development arena, to achieve a more comprehensive and innovative understanding of complex development situations. In evaluating network performance, it is therefore appropriate to pay special attention to the quality of the learning process.

Evaluating Network Performance

The value added by networking
What is the value added to development through networking? What impact does it have on NGO performance? And how can we measure it? Does networking have a direct impact of its own, or only an indirect one, through improving the impact of network members? We are as yet far from being able to answer such questions convincingly. The

study of networks among NGOs, and their effects on the work they do, has only just begun. Hence, opinions vary widely between supporters and critics of networking.

To be able to answer such questions more systematically, the first thing we ought to do is to set a standard, to state what we expect networks to contribute. This is what all networks do for themselves, but no universally applicable formula has evolved so far. What follows is my contribution as a first step in this direction. This is, however, only a first attempt, a contribution to the discussion rather than the definition of a detailed standard at this early stage.

As we have seen, NGOs and their leaders are motivated to network because it helps them to improve their operations. If we take this as a point of departure, we may look at networks of NGOs as 'quality circles', designed and operated to sustain and raise the quality of our work, outputs and impact. This is exactly what networks ought to be. Networks are successful when they help us improve our performance. If they do not, they collapse under the pressure of our other day-to-day obligations. Such a contribution to performance can be of a temporary or a permanent nature. So not all networking activities become permanent or institutionalized. Yet, if they do, it is because those investing in the network expect its contribution to performance to continue.

Following the analysis presented above, networking efforts can be evaluated against their contribution to the three U's, as follows:

° They may help *upgrade* the quality of the activities, outputs and impact of member NGOs, by providing mutual support and services on the basis of a joint assessment of needs.

° They may facilitate a collective learning process among their members, helping to move the analysis of development problems *upstream*.

° Networks may contribute to an *upshift* of NGO activity, redirecting it towards national and international audiences.

Finally, networks may also be expected to incur costs for developing, administering and evaluating networking activities. These are the only overhead costs associated with network operations per se.

There is no reason why we should not try to be as rigorous in evaluating network performance as we are in evaluating NGO performance in general. Donors rightly expect any network they fund to specify its expected outputs and impact, and to define indicators for measuring these. However, at our more general level in this paper we will focus mainly on the nature of networks as instruments for learning together, for helping NGOs in the permanent reformulation and adaptation of their role with respect both to the rural poor and to government institutions. Taking the above criteria as a starting point, I feel we may be able to do just that.

A framework for self-evaluation

As a first attempt, we may formulate a framework for evaluating network performance and impact, as summarized in Table 1. The table presents possible indicators for assessing network achievements for each of the four criteria outlined above. I suggest retaining a distinction between performance and impact, the latter word being reserved for the ultimate impact of the network on the lives of its members' clients. Performance, then, means the contribution the network may make towards improving the effectiveness, efficacy and efficiency of members, or in other words their ability to achieve an impact. The performance of the network is to be the prime focus of evaluation, not its ultimate impact. This must be so because the latter is hard to measure, as a Dutch NGO impact study illustrated recently (de Wilde et al, 1991). The difficulty lies in separating the effects of networking on impact from the effects of members' other activities. In addition, as we have seen, it is the quality of the learning process created by the network among its members that determines to a considerable degree whether the network will succeed or not.

In formulating both performance and impact indicators I have used the terms efficacy, efficiency and effectiveness (Checkland and Scholes, 1991). Efficacy refers to whether the means we use (i.e. functional groups, savings programmes) actually work in producing the desired effect (i.e. use of improved farm technologies, better health care, or increased savings among the poor). Efficiency refers to whether the same results could have been achieved with fewer inputs, in other words the ratio of outputs to inputs. Effectiveness refers to whether our efforts and outputs have in fact contributed to achieving our longer term aim (i.e. the eradication of poverty).

In the following paragraphs I will briefly consider the main indicators outlined in Table1, saying why I feel them to be relevant.

Evaluating services

One of the prime functions of networks is to make optimum use of the resources available within member organizations to strengthen the performance and impact of other members. A joint assessment of strengths and weaknesses is therefore one of the first outputs we may expect of a network. The existence of a systematic needs assessment and a resource inventory (i.e. expertise and facilities), and their quality, may therefore be taken as an indicator of network performance. Obviously, both will need regular updating. The degree to which they are up-to-date and the frequency with which they are used by members will be important indicators of their quality.

The quality of the services provided by members of the network to others, or by the network organization to its members, should be the focus of continuous attention, adaptation and refinement. The evaluation may assess the 'closeness-of-fit' of services with the mission of the network and the needs articulated by network members, the frequency with which members make use of services, and the content of the services themselves.

A tricky issue in the provision of network services is the allocation of costs. Although the benefits of services can be clearly located with individual member NGOs, deciding whether to make individual members pay for services is difficult. It is not simply a matter

Table 1 A possible framework for evaluating NGO network performance

A. Main objective B. Main function	Network performance indicators	Network impact indicators
A. Upgrade NGO performance B. Services	• Quality of resource inventory and needs assessment • Closeness-of-fit of services with mission • Quality of services • Intensity of use of services by members • Allocation of costs	• Total change in efficacy and efficiency of members
A. Move NGO activities upstream B. Learning together	• Quality of joint learning processes • Coverage/distribution of learning experiences • Definition and transparency of technical and methodological standards • Clarity of analysis of development issues	• Total change in efficacy and effectiveness of members
A. Create upshift in NGO activities B. Advocacy	• Frequence and relevance of external contacts • Articulation of alternative development proposals • Increase in members' participation in public development debate	• Total increase in members' impact on development policy and debate
A. Network development and maintenance B. Network management	• Roles of different network actors in developing the network's mission and organizational plan • Relevance of participating NGOs to network's purpose • Design and operation of network communications infrastructure • Design and operation of financial and administrative structures • Quality of decision-making procedures • Efficacy and efficiency of secretariat or facilitation unit(s)	• Effectiveness of network operations

of money, but also of time and good will on the part of the service provider. Moreover, the people and organizations who most need a given service may be the ones least able to pay for it. Within each network, therefore, decisions must be reached as to which services should be paid for through a general (membership) charge and which should be directly charged to users only. Evaluations should address both types of services, and the ways in which decisions on cost allocation were reached. A service that is paid for by users yet much in demand even by the least affluent members of the network is clearly a good service.

Evaluating learning together

Learning together lies at the heart of networking, yet it is the most difficult activity to evaluate. Most network members can tell you whether the network provides them with new ideas, stimulates them to learn and to try out new practices, but they will be hard-pressed to put their finger on exactly how it does so. Still, given the importance of this activity we will have to try to find appropriate indicators.

To evaluate learning together we will have to adopt a qualitative approach, looking at the process rather than seeking only to define the products. One approach could be to look at the settings in which learning takes place (Rap, 1992). How do network members learn from each other? Do they do so by working together, perhaps experimenting together, or by developing a policy document in a task group, by attending a course or workshop, by watching and discussing each other's practices during exchange visits, or by temporarily swapping jobs with each other, or even in a sort of apprenticeship, understudying more experienced staff in other member NGOs? Rap touches on another issue relevant to our discussion here: to what extent are visual, discursive and physical experiences part of our learning settings, or do we mostly concentrate on one of these only? The ancient Chinese distinction between hearing, seeing and doing, and the degree to which we can learn from each, also applies to networks.

An important part of the evaluation of learning together consists of assessing the degree of participation of network members in the learning experiences organized by the network. Is participation distributed evenly among the staff of different members? Does the network actively stimulate wider participation? And, if participation appears limited, is this for logistical or financial reasons, or do the subjects covered in learning experiences simply not relate to the needs of different member organizations as expressed in the needs assessment?

Two questions can be asked which may be considered apt for a more product-oriented evaluation. First, has the learning process led to a clearer definition of the technical and methodological standards to be set for NGO interventions? The degree to which a network achieves consensus in these areas may be one of the best ways of evaluating its performance. Second, have members achieved a better understanding and a clearer analysis of current development issues? Again, this should be an important output of the learning process.

Evaluating advocacy

Several indicators may be useful here. We could begin by looking at the type, relevance and frequency of the external contacts organized by the network. The degree to which the network succeeds in facilitating the formulation of alternative development proposals could also be examined. Lastly, the actual participation of network members in the public debate or in the development of government policies needs to be assessed.

Evaluating management

Networks have to be managed as learning environments, as a space for study and reflection. Management is to be evaluated in its role as facilitating rather than controlling.

The type of facilitation needed during different phases of network development, may differ greatly. To account for this, I distinguish three different phases: (1) planned activism, (2) creating social synergy, and (3) maintenance and development. The first phase, as we have already said, is the one in which the network passes from an ill-defined sense of 'we should do something' to the formulation of a mission and a plan of activities. In the second phase there is more emphasis on facilitating interaction, activities are further developed and a communications infrastructure is created. During the third phase, the network has reached a certain maturity.

During the planned activism phase, network start-up is the key management objective. Evaluation may look both at the process and at its results or outputs. On the process side we may consider the role of participants in developing the network's mission and plan of activities. Broad, effective participation of (prospective) members is essential here. Only in this way will they acquire a sense of ownership of the network.

However, a network probably cannot come into being without a motor, that is to say an active secretariat or facilitating unit. That is why, in evaluating network experiences, the role of the network secretariat almost always emerges as an important theme. In the view of ALIN's organizers, it is one of the crucial issues. How much initiative should the secretariat take? How much leadership should the network staff show? Echoing the ALIN evaluation team, one might ask: is a network anything more than an information service run by a secretariat? In my view, we should not try to set rigid standards for participation during the early stages of network building, but we should study carefully the level of participation achieved and compare this with the available means and intentions of the network's prime movers. At a more general level the roles of different actors may be assessed, including prime movers, sponsors, facilitators and prospective members, along with that of the secretariat. What contributions did each of them make, in terms of time, energy, ideas and, last but not least, funds?

On the output side, important issues for evaluating management during the planned activism phase are the procedures which have evolved for taking decisions within the network. How effective are these in achieving consensus on important issues? In addition, the clarity and focus of the network's mission statement can be studied. And we may also consider whether the NGOs included in activities at this stage are those deemed relevant to fulfilling the network's mission. Finally, the proposed outlines for the organization of the network, its finances, and its communication infrastructure can be studied, so as to gauge the progress made in building the network.

In the social synergy phase the participants build on the foundations laid during the planned activism phase. A more complex web of interactions between members evolves. Pilot programmes and activities are launched, and special problems are addressed through task groups. Crises are part of this phase. Evaluators should be aware of the role conflicts play in building up a healthy organization. Again, participation is fundamental, as is the effectiveness of procedures for achieving consensus.

During this phase probably one of the most interesting issues for evaluators, and for managers too, is the degree to which the management style of the secretariat and other facilitators stimulates or suffocates innovation and learning within the network. Those involved may wish to consult recent management literature on innovative or learning organizations. Decentralizing decision-making, stimulating broad participation, allowing space for dissent and managing conflicts are recurrent themes.

In the maintenance and development phase the core activities of the network are well established, its services operational, and its mandate and decision-making procedures well defined. During this phase, outputs and results can be described in more detail, and monitoring and evaluation become less qualitative and more quantitative. Quantitative indicators revolve essentially around the costs and benefits of the network, seeking to specify monetary and other outlays, and benefits in terms of impact on the work of member NGOs.

Understanding network impact

While the conclusions we may draw with respect to the efficacy and efficiency of networks are necessarily limited, this is even more true of their effectiveness. Evaluating impact would mean comparing the network's mission with the wider needs of society. How does 'networking for sustainability' help to bring about more sustainable forms of agriculture? We normally reply to such questions in the negative: without networking, what chance would sustainable agriculture have? It seems obvious to most of us, but this sort of answer will not convince the sceptics, who ask: aren't there better ways of achieving sustainable development than these endless conferences, workshops and discussions you people have? Even to people like me, who believe in the value of networking, it is obvious that the process needs a lot more thought. Social research into the effectiveness of networking should, I suggest, be a priority.

Conclusions

Networking among NGOs has increased over the past decade, particularly among NGOs active in the field of sustainable agricultural development. From the experiences presented in the papers that follow this one, we can easily see why: sustainable agricultural development requires a level of action and reflection beyond the powers of any single development organization.

Increasingly, too, networking experiences are documented, reviewed and analyzed from the point of view of performance. The papers in this book are but a few examples. It is a sign to those responsible for NGO operations and funding to take networking seriously. Networking is a valuable tool in the kit NGOs have at their disposal. Moreover, networks can be understood, facilitated, managed and evaluated systematically—although their full implications are not known to any of us yet.

A major effort is needed to develop better ways of designing, managing and evaluating networking activities. The evaluation process, in particular, needs to take into account the special characteristics of networks, which set them apart from other forms of human cooperation. The simple transplantation of monitoring and evaluation models designed for conventional organizations is not enough. Some interesting approaches to this challenge are found in Checkland and Scholes (1991). These have been further developed for sustainable agriculture and natural resource management by Bawden and his colleagues at the University of Hawkesbury, Australia (described in Wilson and Morren, 1990).

Finally, serious research into networking and network management may help to solve some of the critical issues which remain unclear: how pro-active can the secretariat be without suffocating the network? How formal must a network become to be permanent? At what levels should networks intervene, and how should networks operating at different levels relate to each other? What is the ideal structure of a network? What are the consequences of networking for the relationships of NGOs with other parties, notably the grass roots organizations they are working with? How should costs be allocated to meet efficiency criteria on the one hand and participation criteria on the other?

As I see it, research on networking will have to be on our priority list for some time to come.

References

Bourdieux, P. 1991. *Language and symbolic power*. Cambridge: Polity Press.
Checkland, P. and Scholes, J. 1991. *Soft systems methodology in action*. London: John Wiley and Sons.
El Taller. 1990. Think globally, act locally and... act globally! A challenge for NGDOs in the 1990s. Report on first conference of the think–tank. Reus, Spain.
Engel, P.G.H. 1990. Knowledge management in agriculture: Building upon diversity. *Knowledge in Society: The International Journal of Knowledge Transfer* 3 (3): 28-35.
Haverkort, B. and Ducommun, G. 1990. Synergy and strength through networking. In: *ILEIA Newsletter* 6 (3): 28-30.

Rap, E. 1992. Learning in practice: Farmers' learning processes in agriculture. Wageningen Agricultural University, unpublished M.Sc. thesis.

Tajaroensuk, S., van Reisen, M., Schutte, T. and Zwanborn, M. 1992. *International human rights organisations*, Chapter 6.1: Findings on human rights information and documentation system (Huridocs). DGIS/NOVIB, The Hague.

Wilson, K. and Morren, G.E.B. (eds). 1990. *Systems approaches for improvement in agriculture and resource management*. Macmillan, New York.

Address

Paul Engel, Wageningen Agricultural University, Department of Communication and Innovation Studies, P.O. Box 8130, 6700 EW, Wageningen, Netherlands.

Networking for Sustainable Agriculture in Peru: Experiences of the Red de Agricultura Ecológica

Enrique Kolmans

Introduction

The Red de Agricultura Ecológica (RAE) is a network in Peru consisting mostly of non-government organizations (NGOs) promoting environmentally sound agriculture at the *campesino* level. RAE seeks to cover the whole country and there are now members in almost all regions. Six regional groups exist, active to greater or lesser degrees.

The network has approximately 50 members. Organizations are full members; individuals have non-voting status. The greatest interest and the most active participation has been shown mainly by the heads of field programmes and by extension staff. Government and university members participate actively, as well as innovative farmers. There are a few expatriate advisors. Farmers' organizations and rural communities have increased their representation lately.

During the first few years the network was semi-formal, but members have now asked for better membership registration and regulations.

Coordination of the network is in the hands of a coordinating group called CONAE (the National Coordinating Group for Ecological Agriculture). This consists of five well established NGOs with headquarters in the capital, Lima, one of which provides the network with an office and other services. Since the middle of 1991 the network employs two full-time staff, an economist and an agronomist.

In its present form the network concentrates its activities in four areas:
- Coordination of members' programmes
- Training and development of training materials
- Exchange of information
- Influencing public opinion.

This paper describes why the network was initiated and how it has developed to become an important tool in promoting environmentally sound agriculture in Peru. It also outlines the main lessons from the trial-and-error process we went through in developing the network, so as to increase understanding of the strengths and weaknesses of the networking approach.

Network Development

The idea

During the 1980s several Peruvian NGOs had worked on promoting environmentally sound agriculture and had met with considerable success. The positive experiences of the Huertos Integrales Familiares y Comunales (HIFCO) in Pucallpa, of the Centro Ideas (IDEAS) in Cajamarca and of the Instituto de Desarrollo y Medio Ambiente (IDMA) in Huánuco, as well as the growing interest and expectations of their foreign partners, led in 1989 to an initiative to organize a national meeting on environmentally sound agriculture in Peru, with a view to promoting this approach to rural development more widely throughout the country. The approach seemed especially important for the small-scale farming sector, in which about 200 Peruvian NGOs were known to be working. It was felt that these organizations together could have an important influence on the course of rural development in Peru. At that time most of them were working with conventional agricultural technologies and approaches.

The first national conference on environmentally sound agriculture was held in Lima in August 1989 and was attended by 29 NGOs and several representatives of universities and other institutions. The organizers hoped that the meeting would stimulate closer cooperation among NGOs and between them and other parties. Interest and participation in the meeting was unexpectly great, and the idea of increased cooperation met with general approval.

The initial objectives

At the Lima conference it was decided to pursue and develop the idea of networking. The following goal was formulated:
- To promote environmentally sound agriculture as part of an eco-development strategy based on a balanced relationship between nature and society, through the sustainable use of resources without incurring technical and economic dependence.

An environmentally sound approach to agriculture was defined as an approach which:
- Conserves and enhances natural soil fertility to improve crop yields and the living conditions of farm families.
- Avoids the use of agrochemicals.
- Identifies and applies positive technical components (traditional and modern) which meet social, economic and environmental requirements and which ensure the production of enough food.

The participants at the first national conference agreed to form a small coordinating group to support other organizations working on environmentally sound agriculture and to begin networking with and among them. This coordinating group was given the tasks of organizing the exchange of information, developing guidelines and criteria on technical and scientific aspects, assisting members in planning and executing projects, organizing training for technicians and extension workers, and promoting environmentally sound agriculture through the media. It consisted of five member organizations, namely IDEAS (chair), IDMA, the Centro de Asistencia en Proyectos y Estudios Rurales (CAPER), Diaconia, and the Fundación Peruana para la Conservación de la Naturaleza (FPCN), all with offices in Lima. It was supported by an advisory council consisting mainly of members of the agricultural university in Lima. Until 1991 CONAE worked without any specifically assigned staff or financial resources.

The network develops
The development of the network has been marked by three highlights: the national meetings of 1989, 1990 and 1992. These meetings played a decisive role in determining the direction the network has taken.

After the first conference in 1989 the coordinating group, CONAE, started its work. Leading up to the second national conference, held in Cajamarca in August 1990, it concentrated its efforts in the following four areas:
° Preparing network papers for discussion and approval at the second national meeting. These included a paper on the basic characteristics of environmentally sound agriculture and its role and importance in Peru. A second paper suggested guidelines for experimental work and the documentation of experiences. A proposal for short training courses for technicians and extension workers was written. Another project proposal outlined ways of increasing public awareness of the need for environmentally sound agriculture in Peru.
° Giving lectures and participating in fora held in different parts of the country on environmentally sound agriculture, particularly the problems associated with the use of agrochemicals.
° Maintaining contacts with like-minded organizations elsewhere in Latin America. These included the Comisión Coordinadora de Tecnología Andina (CCTA) and the National Pesticide Network in Peru, and the International Federation of Organic-Agriculture Movements (IFOAM) and the Consorcio Latinoamericano de Agroecológica y Desarallo (CLADES).
° Publishing the proceedings of the first national meeting.

The participation of about 60 organizations at the second conference in Cajamarca exceeded all expectations. The conference focused both on content issues, such as a comparison of different approaches to agricultural technology development, and on administrative issues, including networking at different levels (national, regional, international). Most important, discussions took place at sub-workshops on the project proposals prepared by the coordinating group. This conference launched the network as a formal entity. Six semi-autonomous regional groups were established. IDEAS, IDMA and Diaconia were confirmed as members of CONAE, with the Centro para el Desarrollo Regional (RAIZ), the Centro de Estudios y Promoción del Desarrollo (DESCO) and the Centro de Investigación, Educación y Desarrollo (CIED) becoming new members.

Regional groups define activities

The action plan for 1991-92, defined this time with a strong input from regional delegates, concentrated on strengthening the movement for environmentally sound agriculture within the regions. Provision of relevant information and training was given highest priority. Activities to be carried out before the third conference in 1992 included:

- A survey, based on a questionaire, to find out members' actual involvement in environmentally sound farming.
- Setting up of a library service offering copies of publications on environmentally sound agriculture to members. Today this library also provides a question-and-answer service.
- Publishing of the network's newsletter, *Cultivando: La Red de Agricultura Ecológica*.
- Publishing of the proceedings of the second national conference.
- Coordination with regional representatives, promoting more activity in the regions.
- Preparation of a 1-week introductory course for technicians and extension workers on the principles of environmentally sound agriculture, including training materials.
- Holding of six training courses on aspects of environmentally sound agriculture in different regions. In fact the demand for such training is much greater than this, but time and finances do not allow the training team to meet it all.
- Creation of task forces for topics such as: clarification of concepts and work strategies; production of and trade in organic food; education and research; economic feasibility of environmentally sound agriculture. The work of the first two task forces led to a workshop and a seminar respectively on their topics.
- Fund raising, so as to employ a small permanent staff for training and advisory tasks.
- Evaluation of further fund raising potential and development of project proposals.
- Application for scholarships for courses abroad and selection of participants from the network. Unfortunately, in spite of great effort put into this, success has been limited, with the exception of scholarships for a few seminars held by CLADES. The network now puts most energy into setting up training facilities in the region which can be shared with other Latin American networks.

The top five priorities now

The third national meeting was held in Huánuco (Central Peru), which is situated, like Cajamarca, in the Andes, but in a zone which has been affected considerably by political violence. Although it was risky to hold it there, IDMA's important practical experience in the area made the risk worthwhile. Compared with the former national meetings, which had a rather theoretical flavour, the Huánuco meeting centered on the practical application of environmentally sound agriculture. The 19 participating NGOs had been invited because of their concrete achievements and experiences in the field, and came from all parts of the country.

Based on the analysis of the different experiences presented and the problems encountered, the participants identified the five top priorities for the further development of environmentally sound agriculture.

First of all people agreed that it was of the utmost importance to document successful experiences. Second, it was felt that the training courses should be continued and intensified. Third, there was a need to improve the methods for on-farm research with farmers, which currently follow an approach which is too conventional and top-down. Fourth, marketing and trade issues, as well as the issue of economic feasibility, needed much more attention. And finally, improving existing data bases was crucial if networking opportunities were to be enhanced.

Once again the conference grappled with the question of how to enhance participation in the network of members from the regions. But the example of the Cuzco regional group showed that a decentralized approach could be effective. Other organizational issues were also dealt with. The five members of the national coordination group were re-elected.

The Huánuco conference, with its emphasis on involvement and activation of the regional groups, led to a real gain in the quality and quantity of work by network members, although several problems have yet to be solved.

Networking within Latin America

The activities of this network in Peru should be understood in the context of emerging network activities within Latin America as a whole. Our own network derived some of its momentum from a meeting held in Cochabamba, Bolivia, in 1989, attended by representatives of 60 organizations. The participants decided to build a movement dedicated to environmentally sound agriculture and sustainable rural development and to establish it as the Latin American branch of the International Federation of Organic Agriculture Movements (IFOAM-LA). Since Cochabamba, one of IFOAM-LA's main priorities has been to encourage the formation of national networks similar to our own. Drawing on the Peruvian experience, considerable progress in networking has been made in Brazil, Argentina, Chile, Paraguay, Uruguay, Mexico and Central America. Of greatest importance for our network in Peru are developments in other Andean countries. Close links have been established with national networks similar to ours in Ecuador and Colombia, especially in the field of training.

Last year IFOAM-LA became the Movimiento Agroecológica Latinoamericano (MAELA), as many members of national networks were not members of, nor had links with, IFOAM. Despite its new name the organization remained committed to coordination and furtherance of effective decentralized networking in each country.

Two members of the CONAE group, IDEAS and CIED, are also members of another organization linking people in different countries of the continent, namely CLADES. This is a group of 12 associates in different Latin American countries. It does not consider itself a network or movement. Its main aim is to strengthen the individual associates so that they can offer (together or each in its area of influence) more specialized assistance. It publishes a well presented newsletter aimed at universities and government agencies, as well as NGOs. As CLADES considers the strengthening of national networks an important issue and there are direct personal links between CLADES and CONAE members, constructive collaboration has evolved. But in Latin America as a whole the implementation of the CLADES approach seems to have lost its momentum.

Organic production and trade
The production of and trade in organic products is an issue that has come high on the network's agenda. Whereas local and international markets for conventionally grown crops are frequently saturated, organic produce could prove a lucrative option for smallholders. The increasing demand for certified organic cereals, fruits, vegetables, livestock products, herbs, health products, fibres and textiles, with the considerably higher prices that they command, is an opportunity not to be missed.

Small-scale agriculture and livestock husbandry in Peru is still practised mostly without the use of agro-chemicals and in areas free of environmental pollution. And because of the special agro-ecological conditions under which they are grown, several crops, including quinua, amaranth, some legumes, coffee and herbs, are renowned for their taste and/or their high nutritional value. Traders have shown an interest in buying these products, provided they are certified fully organic.

This is a real challenge for our network. The progress already made by some members has encouraged the network to take initiatives in this area. A seminar on organic production and trade took place in March 1992. As a follow-up we are working on the development of production and trade standards and providing direct assistance to members in this field. Success on this issue would contribute greatly to increasing the economic viability of environmentally sound agriculture in Peru.

Achievements and Disappointments

It is too early yet to assess the impact of the network. But the experiences of the first few years are very encouraging indeed. Among our main achievements the following seem the most important:
- People committed to environmentally sound agriculture have been identified throughout Peru and have come to know each other.
- A strong coordination group has been established.
- A decentralized, participatory network has taken shape and is functioning well under Peru's sometimes difficult conditions.
- Work strategies and guidelines for network members have been developed.
- A permanent information and training service has been established.

If one analyzes this success, the most crucial factor explaining it is probably the initial commitment of a significantly large group of people. From the beginning the network was not the project of just a few people only. Although the network has developed by trial and error, sometimes with 'spontaneous' decisions leading to unrealistic goals, the general direction taken has maintained its integrity.

One example of unrealistic goals was the priority given in the first years to raising awareness within Peru of the need for and the potential of environmentally sound agriculture. This requires more evidence than was available at that stage.

In retrospect, the time spent during the first year, especially by the Lima-based coordinating group, on discussions and documentation of the theory of environmentally sound agriculture and of 'correct' research methods had little impact on the network and its development. These discussions were not relevant for the NGOs working with the *campesinos*, although they may have been so to some consultants and individuals within CONAE. The updating of records and documentation of members' achievements, together with training activities, would have been more in line with members' needs and might have produced results more quickly, had they been initiated earlier.

A major difficulty in developing the network was the one-sided belief that everything in traditional Andean agriculture is environmentally sound and/or sustainable. Although traditional systems have many outstanding features (some of which are of inestimable value), this narrow point of view does not take into account the pressures to which traditional systems have long been subject and the declining efficiency in the use of natural resources in such systems, mainly resulting from more than four centuries of exploitation and subjugation by European conquerors. Unfortunately, the attitude of some supporting circles in the industrialized countries, unaware of the realities of daily life in the Peruvian Andes and of the aspirations and needs of the rural inhabitants, did little to dispel this illusion. The challenge is to identify positive technological elements from both traditional *and modern* agriculture, and to combine these in ways which meet *social and economic*, as well as environmental, criteria. The long discussions on this issue within the network finally brought this home to most members and supporters.

During the past 2 years the network has shown greater flexibility and has launched semi-autonomous regional groups, decentralizing at least some activities to these regions. Without this change there was a real risk that the coordinating group would start to play too central and dominant a role, as has happened elsewhere. It was understandable that the members from the regions expected the CONAE group, with its greater experience and knowledge, to play the role it played at the beginning. The members also felt the need for guidelines and training, which had to be organized and delivered by a group and individuals with certain capabilities. Yet the cultural and physical distance from Lima to rural reality often made these centrally organized inputs seem unrealistic and too theoretical to those in the field.

Another weakness in our networking approach is the lack of farmer involvement. This needs much more attention in future. The experiences of the Campesino a Campesino Movement in Mexico and Central America should provide insights as to how to encourage a similar approach in the Peruvian context. Experiences of farmer networks in Asia, as presented at a recent international workshop on this subject held at Silang, the

Philippines, may also be relevant. A *campesino* section within our network, or an autonomous farmer-to-farmer network in close cooperation with our own, needs to be developed in the near future.

One should not forget that this network has evolved during one of the most troubled periods in Peru's history, in which rural development work has been particulary difficult and dangerous.

Conclusion

The RAE has been useful, especially at a time when many NGOs are experiencing teething problems and struggling for greater efficiency and effectiveness through cooperation and the exchange of experience. It demonstrates that networking is possible without a great deal of financial support from the outside, using members' contributions in kind as the main source of capital.

Address
Enrique Kolmans, Friedrichstrasse 13, W–7143 Vaihingen/Enz, Germany.

The Association of Church Development Projects in Northern Ghana

Malex Alebikiya

Introduction

The Association of Church Development Projects (ACDEP) was formed to facilitate the exchange of experiences and initiate joint activities and support services for the benefit of church-based development projects in Northern Ghana. This paper traces the evolution of ACDEP, examines the constraints it was set up to overcome, its objectives, role and activities, and its effect in building links between different projects and with government services.

Origins

The first church-based projects in Northern Ghana emerged during the 1950s and 1960s. They grew rapidly in number through the 1970s, reaching 23 projects by 1980. ACDEP now consists of 24 church-based projects, including 15 church-sponsored health projects and 2 training projects sponsored by Canadian University Services Overseas (CUSO), in addition to 7 projects devoted to agriculture. The main church bodies responsible for promoting and administering these development programmes were the Committee on Churches Participation in Development (CCPD) and the development offices of the country's three Catholic dioceses. The CCPD (formerly called the Christian Service Committee, CSC) is the development arm of the Christian Council of Ghana, which is made up of the country's Protestant Churches. The CCPD is officially responsible for some of the projects of the member churches, while others are coordinated directly by the sponsoring churches. The role of the CCPD includes both coordination of ongoing project work and assistance to the development of new projects in response to community needs. The development offices of the three Catholic Churches have similar functions, but their work is restricted to the individual dioceses.

This complex church-based structure had several weaknesses:
° The projects operated in isolation. The only organizational relationship they had was with their sponsoring church. In the absence of any horizontal coordination between projects, agencies tended to replicate programmes and operate similar activities within the same locality. For example, the Saboba Farming Cooperative Society and the Saboba

Evangelical Protestant Church Agricultural Station carried out similar functions within the same community.
- Coordination based on the vertical administrative structure of the church and parish was non-specialist. Not being agriculturalists, coordinators were unable to provide the technical back-up required for initiating, planning and implementing projects. The major emphasis of project management was on the submission of annual and financial reports, the preparation of budgets, the regularization of reporting procedures and other such administrative activities.
- There was no technical training for the largely expatriate staff to undertake development work in rural communities.
- On the ground there existed different development approaches, without the benefit of a learning process between projects.
- Policies regarding the delivery of services were conflicting and tended to undermine each other. For example, while some projects promoted animal traction others developed cooperative tractor services, sometimes in the very same community. Again, while some encouraged farmers to use organic manures, others were active in the sale and distribution of fertilizers.
- The lack of coordination among projects led to the creation of parallel and sometimes inefficient and costly services. Both CCPD (then CSC) and the Catholic Archdiocese of Tamale launched separate input supply projects, selling the same inputs at different prices.
- There was no single voice on agricultural development policy. Hence the projects introduced new components piecemeal rather than adopting an integrated approach. Because the projects worked in remote communities, government was content to have them fill the gaps.

In 1977 the Church Agricultural Project (CAP) was formed. It was a loose association of agricultural project managers and technical staff which aimed at promoting collaboration between the various projects and facilitating the exchange of information. In 1978 a similar association of family health projects was formed. The two associations worked closely together, deliberating on common problems. In 1988 the two associations merged to form ACDEP.

Description of the Association

Objectives
The objectives of ACDEP include:
- Bringing all church projects together to share experiences.
- Improving horizontal communication between projects on technical and management issues, and sharing information and experiences.

- Procuring and delivering inputs to agricultural projects more efficiently.
- Organizing education tours, symposia, workshops and other events.
- Fulfilling a social need for interaction among the technical (non-clerical) staff of the churches. This was seen as particularly important as the agricultural staff felt isolated in remote rural communities.

Behind all this was the assumption that the church projects, as non-government organizations (NGOs), had a distinct philosophy and motivation that made them different from government extension programmes. They were expected to have a stronger emphasis on equity, concentrating on the needs of the small, resource-poor farmer.

Status and roles

ACDEP is recognized by the churches as a professional association of managers and technical staff whose main concern is the technical quality of their programmes and the efficient delivery of services at the project level. In this role the association is seen as complementing the functions of the CCPD and the Diocesan Development Coordination office, whose main concern is with administrative efficiency. The churches have fully supported the development of ACDEP. They have also sponsored activities and projects proposed by ACDEP.

ACDEP plays an important role in initiating and coordinating specialized support services for the church-based projects. These include:
- The Agricultural Information Service (AIS), which focuses on training technical staff.
- The Small-scale Agricultural Extension Support Unit, which develops extension methodologies and approaches suited to resource-poor farmers.
- The Church Agricultural Input Supply Project, which procures inputs for timely distribution to the various projects.
- The Family Health Advisor in Ghana Project, which trains health staff and assists projects in implementing and monitoring their activities.
- The CUSO-CCPD Training Programme, which provides training for field staff and project managers in planning, management, evaluation, community group organization, the role of facilitators, women's participation and participatory methods.

Although not officially registered as an NGO, ACDEP is recognized by government agencies in the region as the mouthpiece of the churches on development issues. Many agencies have links with it and attend its general meetings. For government agencies, particularly the Ministry of Agriculture, it is an important link for exchanging information and discussing agricultural programmes and policies. ACDEP has thus become an important institution for both church and government, as well as for individual member projects.

Structure

As a loose association, ACDEP has an implementing executive elected from among its membership whose duty is to carry out the decisions taken by the association. The

Agricultural Information Service, which implements most of the educational programmes of ACDEP, provides it with secretarial support for writing minutes, typing correspondence and organizing the general meetings.

Activities
The activities of ACDEP in the field of agriculture are inseparably linked with those of the Agricultural Information Service, which is responsible for implementing them. They include:
- Education and training on development issues
- Organizing general meetings
- Identifying and initiating support services to ACDEP members and others
- Linking with government institutions and other NGOs
- Advocacy.

Education and training
This is probably the most important activity of ACDEP. Its objective is to raise the awareness of all project staff on contemporary development issues, and to increase the technical knowledge and skills of field staff.

The first formal in-service training organized by ACDEP was in 1979, when training was provided for extension staff of member projects. Since then a 2-week residential training programme has been organized annually with financial support from the Oxford Committee for Famine Relief (OXFAM) and technical inputs from the Institute of Field Communication and Agricultural Training, an organization originally established to provide in-service training to field staff of the Ministry of Agriculture under the Upper Region Agricultural Development Programme. The scope of the ACDEP training programme and the number of trainees has expanded steadily as the association has gained experience and as the impact of training has become more evident. By 1986 the association was training up to 100 field staff annually on crop and animal production, animal traction, soil conservation, use of fertilizers, extension communication and animation methods, report writing, and other subjects. In 1982 the need for a parallel 2-week training programme for project managers was expressed. This began in 1983. The subjects covered related mainly to project planning and management skills and to agricultural extension methods.

To improve the effectiveness of the residential training, the association organized follow-up training. With the assistance of professional trainers from the Institute of Field Communication and Agricultural Training, a programme was drawn up in which the Institute's trainers visited each ACDEP project two or three times a year to monitor impact, revise relevant sections of the courses and identify new areas for training.

This initial training programme for extension staff and project managers firmly established the need for a long-term in-service training programme with a full-time training coordinator. A programme of this kind was established in 1988, sponsored jointly by CUSO and CCPD and financed by the Nederlandse Organisatie voor Internationale Betrekkingen (NOVIB).

ACDEP meetings are preceded by a major workshop on contemporary development issues (environment, agroforestry, rural poverty) or topics of technical interest (grain storage, on-farm research, participatory technology development, etc). These workshops usually last between half a day and 1 week. Other workshops are also held, independently of general meetings. Many of these events are attended by government personnel and the staff of other NGOs besides ACDEP. It is these regular workshops and seminars that account for the increasing membership of the association.

Meetings

ACDEP holds three meetings a year, in February, July and November. The secretariat (AIS) does all the paperwork. A host project, selected at a previous meeting, organizes board and lodging and the conference facilities. Like all the operational costs of the association, those of meetings are met by members through registration and membership dues. Each meeting attracts between 40 and 50 participants from government and other NGOs in the region.

Support services

By identifying and establishing relevant support services for all church-based agricultural projects, ACDEP has created horizontal links between projects and helped to break down denominational barriers and conflicts, paving the way for better cooperation and more effective development work. A radio link between projects now enhances horizontal communication.

Institutional links

The improved links with other institutions offered through the association serve not only to disseminate the work of the church projects but also to enable them to tap government resources and take advantage of opportunities that would otherwise have been missed. Through the AIS the work of the projects is made known to government, other NGOs and the general public. This makes it possible to identify and invite other relevant institutions to ACDEP meetings, educate ACDEP members on the role of such institutions and identify potential areas for cooperation or sharing.

In this way the association has established good links with the Ministry of Agriculture, the Ministry of Health, and various agricultural and rural development projects, such as the Northern Region Integrated Project (NORRIP).

In 1989 ACDEP organized a workshop on participatory technology development (PTD). Project managers from church-based projects and staff from Nyankpala Agricultural Experimental Station (NAES), NORRIP and the Ministry of Agriculture were trained in ways of involving farmers in research. A PTD sub-network was set up within ACDEP and four projects developed PTD programmes as part of their extension work. A collaborative on-farm testing programme has been developed between some of these projects and NAES.

In recent years more and more agencies—government and non-government—have sought links with ACDEP.

Advocacy

ACDEP has capitalized on its growing prestige and its strong links with other organizations to highlight the need for support to community-based development work on behalf of the rural poor, both inside and outside the framework provided by the churches. Churches have been asked to give serious thought to the work of the projects and to development work in general. Several development issues have been brought to the attention of the churches for action.

Notable examples of the influential role of ACDEP are the establishment of the various support services mentioned above and its important part in the design of the CUSO-CCPD Training Programme. The association also organized two workshops to draw the churches' attention to the need for a development policy and approach based on Christian teaching. In 1991 the association played an important role in averting the closure of some projects in the Navrongo-Bolgatanga Diocese because of financial constraints.

The active role of member projects in the promotion of animal traction in the region sparked off efforts by the Ghanaian-German Agricultural Development Programme (GGADP) to establish animal traction training centres. This led in turn to the establishment of the Ghanaian-German Implement Factory. The factory produces animal traction equipment and spares and trains local artisans. Dialogue between the association, the GGADP and the Ministry of Agriculture provided useful inputs to the justification, design, location and implementation of the factory proposal.

ACDEP is cautious in the use of its political influence. Its careful voice has earned the respect of all parties involved in rural development in the region.

Impact

ACDEP has succeeded in improving coordination between church-based projects and in fostering links with formal research centres and government services. It has also striven to improve the technical competence of projects and their responsiveness to the needs of small-scale farmers.

It has broken down denominational barriers, paving the way for better cooperation in the future. Exchange of experience between individual projects has enhanced their development and growth.

ACDEP has put the church projects firmly on the development map of Ghana. By making visible the presence and important role of the churches in development, it has prepared the way for fruitful collaboration with other organizations and with research centres, including NAES.

Through its educational and training activities in collaboration with the AIS, ACDEP has brought about many improvements in the delivery of services at project level. The shift from a technology transfer approach to a more participatory one, from contact farmers to agricultural development groups, from high-input to low-input sustainable agriculture—these are but a few of the changes now evident in some projects. The change of name of the association, from CAP to ACDEP, is evidence of the growing degree of integration now being sought in rural development. Some of these changes have been brought about mainly through ACDEP's awareness-raising workshops, seminars and symposia.

Source
Reprinted from: Wellard, K. and Copestake, J. G. (eds) 1993. Non-governmental organizations and the state in Africa: Rethinking roles in sustainable agriculture development. Routledge, London.

Address
Malex Alebikiya, P.O. Box 9, Wale-Wale N/R, Ghana.

Andeans Unite:
The Birth and Growth of the Andean Council of Ecological Management

Jorge Manrique, Juan A. Palao and Mourik Bueno de Mesquita

Introduction

What is CAME?
The Andean Council of Ecological Management, or CAME, is a network of seven non-government organizations (NGOs) involved in rural development. It maintains direct working relationships with university researchers and policy-makers. Its relationships with farmers and their communities are indirect, through the NGO members.

The main objective of CAME is to strengthen the work of individual NGO members. These have in common their commitment to environmentally sound approaches to agriculture and animal husbandry in Andean settings. CAME hopes to achieve its objective by combining the resources and skills of its members, and by coordinating their approaches, field methods and activities.

Since it began in 1989 CAME has experienced successes and failures, has gone through a deep institutional crisis, and has managed to capitalize on its experiences to develop and strengthen its network activities. In this paper we try to share the many lessons that we feel can be learnt from these experiences.

The need for networking
The CAME network was set up following an analysis of 20 years of experiences in development activities in the Andean region. This analysis showed the limitations of traditional project interventions: the technologies and methodologies used were clearly inadequate to improve life for the majority of the rural population. The NGOs active in the region lacked the experience and support needed to develop alternative approaches such as those known as low-external-input and sustainable agriculture (LEISA) and participatory technology development (PTD). Some small-scale experiences with LEISA and PTD had shown promising results, but due to the lack of institutional coordination and collaboration these could not be developed into real alternative approaches and did not reach a wider range of institutions.

Some NGOs saw the new network initiative as an opportunity to enhance their own interests and broaden their sphere of influence in the region.

The Network Develops

In the course of its history CAME has gone through four stages: start-up, early activities, a crisis and, finally, consolidation and development.

Start-up
The idea of developing a network grew first in the minds of a few individuals, then matured collectively through a 'motivator group'. This group further developed the idea, expressed it in the form of a project proposal and started looking for funds. It also made an inventory of potential members, applying the following criteria: experience in LEISA/PTD, programme stability and managerial capacity. The donor identified was the Netherlands-based Interkerkelijke Coordinatie Commissie (ICCO). Its role was to provide both funds and backstopping. The donor played an important part in clarifying objectives and overcoming the later crisis.

Early activities
CAME began by providing technical and methodological support services to NGOs and organizing exchange programmes. Support activities were on an ad hoc basis, responding to the needs of members in their daily work. They were not based on a common plan. This approach enabled CAME to learn about its members' activities in the field of LEISA, as well as their strengths and weaknesses. It also served to establish links among the NGOs and to find out whether there was in fact a need for a permanent advisory service of the kind it could offer. It was an important period for creating awareness of the need for participatory planning.

Crisis
During its second year the network experienced a deep institutional crisis. This was caused by the top-down attitudes and styles of management and of communications to network members, which prevailed in spite of the acknowledged need for horizontal relationships and cooperation. The initial organizational design and management structure were not appropriate and the members did not understand their role. There was confusion over the functions of the network policy-makers and those of the implementers. Members of the network's Board began interfering in the day-to-day activities of staff, ignoring the Executive Committee supposedly responsible for supervising staff on its behalf.

The fledgling network's limited capacity to manage conflicts, combined with the political interests and personal biases that also came into play, deepened the crisis. Key people within the network began trying to bring it under the influence of their own political party. In Peru's volatile political climate, these moves undermined the basic purposes of the network. Several individuals presented the network as their own creation, in an attempt to further their own careers.

There came a time when the crisis could no longer be ignored. Network activities had to be temporarily suspended. The donor agency requested that differences be resolved and objectives clarified before activities began again. During this difficult period it provided support in the form of a consultancy mission, seeking to make an objective assessment but at the same time expressing its own views and priorities. Finally, the network reaffirmed its commitment to its original idea and organizational structure, and some NGOs and their leaders left the network. The network then went through a 1-year bridging programme to reorganize and plan its field work in more detail.

From this crisis we learned that problems should be brought into the open. All too often one tends to brush them aside to avoid confrontation. By the time it came to a head the crisis almost brought the network to a complete standstill. But facing up to it allowed us to tackle it, to learn from it, and to go on to strengthen our approach. Box 1 lists our five crucial lessons regarding what one should *not* do when starting a network!

Box 1 . *DO NOTs* in starting a network

- Plan network activities from the top (the coordinating group)
- Deny direct participation of NGO members in policy and executive bodies
- Concentrate on implementing (field) activities and forget to allow for the development of the network itself
- Close yourself off from opinion and advice from outsiders
- Accept political, personally biased manipulations.

Consolidation and development

Learning from experience, we initiated a process of participatory planning involving all network members. This resulted in a well formulated 3-year plan integrating proposals from the NGOs into a single CAME proposal. It also led to a change of attitudes on the part of members towards networking and joint activities.

CAME has now become a network with a very specific approach and a quite formal structure, with clearly defined activities. We will now clarify these.

CAME's Characteristics

Choosing an approach

Defining the overall approach of a network involves making choices in terms of its orientation (external or internal), the subject area it focuses on, the geographical area covered and the conditions governing membership (institutional background of members).

Table 1 summarizes the most important factors that, according to CAME's experiences, have to be taken into account.

CAME itself has chosen to network among NGOs working in a limited geographical area, namely the Altiplano of Puño and parts of Tacna and Moquegua, in the Peruvian Andes. This allows us to interact frequently with members and to design activities according to their needs. Internal cohesion and participation are maintained more easily in this type of network.

Of course, this approach has the drawback that CAME's achievements are not automatically shared with organizations elsewhere in Peru, and vice versa, but CAME tries to overcome this by maintaining contacts with other national networks.

CAME's members

The membership of CAME consists of:
- Institutions which have signed the Constitution Act and maintain their representation through specific delegates
- Institutions whose application for membership has been approved by the assembly, following a trial year of formulating and implementing an annual plan of activities relevant to the policies, guidelines and objectives of CAME.

Individuals and government organizations cannot be members of CAME. However, the network establishes working relationships with them through specific agreements.

Members are expected to contribute time, staff, inputs such as experiences and results from field work, and energy for network development. They are obliged to integrate their plans and projects with regard to LEISA and the environment within the structure of CAME before execution. They are expected to contribute to the further development of LEISA/PTD approaches and to further the networking process by assuming specific tasks. In short, if there are no inputs from members there is no networking, no network.

In return, members benefit by gaining access to the experiences and knowledge of others and by obtaining technical and organizational support for their programmes.

Structure

CAME is a formal network. It has legal status and is formally organized. The highest decision-making and policy body is the general assembly of members, which is supported by the Board of Directors. The general assembly chooses from among the members an Executive Committee to coordinate the implementation of the decisions taken and the work of the 'technical team', which acts as the network secretariat. The Executive Committee and the team meet regularly to evaluate activities and adjust their plans. These meetings are a means for exchanging information and coordinating activities.

Presently, the technical team consists of six specialists who together combine expertise in the fields of community development, agricultural engineering, agro-economics, adult education and NGO management. This team provides advisory services, guidance and training for the staff of the NGOs.

Table 1 Considerations in defining the overall approach of an NGO network

Issue	Network approach	Advantages	Limitations
Orientation	Internal	Allows action directly useful to members	Difficult to develop approaches for larger areas
	External	Allows an integrated approach to managing the environment	Stronger dependence on government policy and risk of government control
	Combined	Allows good planning and integral focus; relates members' practices with a wider perspective	Demands greater work efforts and continuous evaluation
Subject of networking	Specialized focus on a single technical subject	Allows rapid advances in tackling concrete problems	Risks separating research from its broader context
	General focus on low-external-input and sustainable agriculture	Allows an integral focus on the agricultural sector	Requires more complex planning and implementation; methods require considerable thought
	Integrated rural development focus	Allows projects and activities to be set in the context of an overall development strategy at local, national and regional level	Requires longer periods, bigger structures and more resources; more difficult to mobilize people
Geographical coverage	Local	Allows site-specific adaptation of technology	Results are not necessarily relevant to large areas
	(Supra-) national	Allows results with an impact for larger areas to be obtained; mobilizes more institutions and donors	Cohesion and participation difficult to maintain; requires more complex working methods
Membership	Limited to NGOs only	Strong common interest of members	Does not allow direct interaction with farmers; can limit technological inputs
	Broader, involving farmers, research organizations, etc	Cross-fertilization between research and development; greater access to new technology, knowledge	Potential for conflicts between research and development objectives

Networking activities such as joint training events, meetings and workshops are dependant on external financing. Members do not contribute funds to these activities. However, they are expected to finance their own local research and development activities, which do not receive funds from the network.

Activities

The activities of CAME span four main programmes:
- Study and documentation in the areas of management of natural resources, sustainable production systems and farmers' organizations. The objective of this programme is to improve the effectiveness of interventions in existing systems and to strengthen farmers' organizations. Analysis and comparison of experiences among both farmers and NGOs is involved.
- Training and human resource development, including analysis of past experiences and development of new approaches. Activities here include seminars, workshops, training courses, working groups and round-table conferences. Tools such as guidelines, interpretation formats, evaluation indicators and planning/monitoring instruments are developed.
- Advisory and consulting services and monitoring. This programme constitutes the support to members provided by the technical team. Activities include participation in members' field work, assistance in the processing of information, support in preparing and implementing courses, organizing workshops, providing information services, preparing project proposals, analyzing experiences, and facilitating exchanges and external contacts.
- Development and management of the network. CAME finds it very important to have a programme specifically devoted to supporting and developing the capacities of members for networking. Important aspects here include: strategic planning, mutual cooperation, policy-making, management and leadership functions, and analysis of socio-economic and political factors that affect the network.

The most crucial elements in CAME's activities are the third and fourth programmes. The consultancy services have a direct impact on the programme and activities of members. They meet a need that cannot otherwise be met. The network development programme help us to pay specific attention to the further strengthening of the network. Without allocating time and money to this aspect it would probably be neglected—possibly leading to the collapse of the network.

Evaluation

Achievements
It is still too early to assess the achievements of CAME, and no detailed analysis will be attempted here. However, there are several areas in which networking has brought significant gains.

The quality of interventions by NGO members and their field staff has been improved through the support activities of the technical team. Through these activities, as well as other internal training and the sharing of experiences, the NGOs have made considerable advances in analyzing and comparing their experiences and results. The first steps in developing appropriate LEISA technologies and tools have been made.

A strong multidisciplinary team has been formed, although it took some time and energy to create real teamwork. Team members are expected not only to have technical skills but also good teaching and communications skills. Such people are not easy to find. Their skills need to be developed through experience on-the-job.

CAME has gained recognition from outsiders. This will facilitate future collaboration. The network has also managed to consolidate its financial position, at least for the first few years.

Limitations

Any network is only as strong as its members. When the programmes and funding arrangements of individual members become unstable, the execution of network activities may be seriously affected.

It has been difficult to compare the experiences of different NGOs and to draw lessons from them because of the lack of methodological instruments and the diversity in attitudes and working styles of NGO members.

There has been too much emphasis during training on building on existing experiences of NGO staff. Analyzing and comparing past experiences was supposed to lead to new insights, and so to improvements in interventions. However, it now appears that training based on past experiences is not always adequate to meet the needs of staff. A more radical approach may be needed.

There is as yet insufficient evidence that the interests of resource-poor farmers and labourers are adequately reflected in the programmes of member NGOs. It is true that these programmes concentrate primarily on LEISA, but they often neglect the fact that the development of technology appropriate to LEISA requires a synthesis of farmers' knowledge with the scientific knowledge of outsiders, not a focus on farmers' knowledge alone. The shift in members' approaches anticipated through interactions in the network is sometimes disappointingly slow to materialize. The network only stimulates and promotes change. It cannot enforce it.

Are the costs justified?

Consolidating a network such as CAME costs time and money. On top of the seed money required for the first feasibility studies come several years of recurrent costs. Tasks such as establishing organizational structures and defining the roles and functions of each level, maintaining advisory services, coordinating the activities of different members and providing administrative services sound mundane but are nonetheless necessary to the smooth running of the network.

Many networks spring up but quickly wither away because their founders did not take these costs into account. Often there is a budget only for field activities and not for developing the network itself. CAME is different, having a fourth programme with its own budget specifically devoted to network development.

Maintaining a permanent consultancy service is especially costly. The main elements here are practical training and the conduct of technical or socio-economic studies on behalf of members. There are also costs involved in developing working tools (checklists, guidelines, frameworks, methods, monitoring systems, etc), providing reference materials and publishing results.

It is very difficult to quantify the benefits obtained through each of these forms of support and, indeed, through all the network's activities. A major challenge facing the network is to find a way of doing so. The central task here is the development of indicators of successes and failures. Only then can we begin to answer the question of whether or not the costs are justified. We believe that, accounting for approximately 7% of the total budgets of its members for field activities, CAME represents a sound investment when compared with normal international consultancy fees.

Lessons learnt

The CAME network was not developed solely as a talking shop. It is based on concrete activities in the field, which it has been set up to service. The needs of each member formed the network's raison d'être.

The agreement that the network should not raise, channel or manage funds for projects or for the activities of individual members avoided its becoming a battlefield for acquiring money.

Management strategies are needed for confronting and overcoming network crises. Opposing interests are generally present during the early phases of a network's development. Bringing conflicts into the open for early resolution would contribute to the growth of the network and avoid more destructive crises later. The role of internal evaluations and external agents is very important in this process.

There is a need to distinguish the policy-making and managerial level of the network from the implementation level. If both levels are mixed up, there will be serious confusion about the role and functions of different parties within the network.

Like other institutions, NGOs are often highly competitive with each other, even engaging in feuds. When grouped in a network they should in theory develop a more cooperative stance, sharing their experiences and results with others. In practice this change seems difficult to achieve. Indeed, to some it seems to appear revolutionary.

Central to the health of a network is the development of a process of participatory planning. This should combine global and regional inputs with those from NGOs and their target groups. This meeting of top-down and bottom-up approaches requires specific methodologies and instruments. A balance must always be sought between the external and internal orientations of the network—between paying attention to relations with other institutions and networks, and to relations among the members of the network itself.

Address
Jorge Manrique, Juan A. Palao and Mourik Bueno de Mesquita, CAME, Jr. Arequipa 128, Puño, Peru.

Networking Among Resource-poor Farmers in South India

Oswald Quintal and Gandhimathi

Introduction

In 1990 a network dedicated to low-external-input and sustainable agriculture (LEISA) was founded in South India. Its members are small-scale resource-poor farmers and non-government organizations (NGOs) searching for alternatives to the current, apparently unsustainable land use practices. All the participants live and work in the semi–arid zones of Tamil Nadu and Pondicherry States. Most farmers here depend on rainfed agriculture and livestock keeping, having little or no access to irrigation. Rainfall fluctuates between 300 and 600 mm per year and soils are of medium to low quality, suffering from nutrient depletion and erosion.

A Natural Resource Base Under Threat

Agriculture in South India, as elsewhere in the world, has evolved slowly. In the traditional farming system the natural resource base was reasonably well protected. The change to modern commercial and chemical farming tipped the balance towards over-exploitation, negatively affecting soil quality, water table, forest resources and genetic diversity. At the same time, most of the resources for agriculture became concentrated in the hands of the few. Currently, about 20% of the population is using 80% of the land. The increasing impoverishment of the majority of the rural population causes near starvation and large-scale migration in search of work, swelling the population of urban areas. Those small-scale farmers remaining on the land face increasing difficulties, having no control over resources such as seed and fertilizers, and being dependent on an increasingly volatile market largely determined by factors beyond their control.

A Network is Born

The need for a network
To improve the lot of these farmers, two approaches needed to be followed:
- Massive afforestation, leading to the regeneration of soil and plant resources.

○ A shift away from the present high-input, commercialized farming system towards one that is more sustainable, ecologically sound and socially just.

In recent years afforestation has gained in importance among government and NGOs working in the region. Efforts on the ground have so far been inadequate and occasionally misguided, but all involved are becoming more aware of forestry issues. The development of alternatives to high-external-input agriculture has received less attention. There have been isolated attempts by motivated farmers and organizations, but the interest of government is still very limited and research began looking at integrated farming systems only very recently.

Against this background, it was felt that NGOs and farmers should take the lead in developing more sustainable forms of agriculture. Most NGOs in the Tamil Nadu area are small. And although they are experienced at working with farming communities on development issues, they often lack technical expertise and have insufficient capacity to develop new technology. Farmers testing new options have done so in isolation from each other and, often, unnoticed by the very NGOs that were aiming to support their efforts. The establishment of a local network combining farmers and NGOs would, it was believed, help to overcome these problems, enhancing the quality of field activities and the speed with which information and technology would be disseminated, and spreading awareness of the importance of LEISA.

The founders of the network

The founders of the network had come together informally for the previous 4 years, discussing and sharing their experiences, before they finally took the initiative of starting the network. Most of them were readers of the *ILEIA Newsletter*. Articles on case studies and networking in other regions, published in this newsletter, had deepened their understanding and conviction. Included in the group were representatives of several NGOs oriented towards social action and agriculture, five environmentally conscious farmers, and a few other committed individuals. One of the founders, the Agriculture, Man and Ecology (AME) programme, had been involved in training NGO staff from South India in environmentally sound agriculture for many years.

The launching workshop

The founding members began activities by making an inventory of farmers and NGOs involved in developing sustainable agriculture in Tamil Nadu and Pondicherry. They then invited these farmers and NGOs to a workshop. In February 1990 the workshop took place, bringing together around 24 NGOs and 7 farmers from the two states to review the need for the network and discuss its objectives and activities.

The following objectives were agreed upon:
• To understand the problems faced by farmers in different areas, in the context of changes that have taken place in agriculture and the related environmental problems.
○ To motivate farmers and organizations to adopt sustainable agriculture.

- To study and document traditional agricultural practices which appear ecologically sound.
- To increase interaction between farmers and organizations and encourage the exchange of experiences, knowledge and skills in sustainable agriculture.
- To disseminate information regarding sustainable agriculture to a wider circle of farmers and organizations.

Activities

The proposed activities of the network, formulated during the first workshop, included:
- Meetings, workshops and seminars to facilitate the exchange of experiences, ideas and skills.
- Study tours to existing 'ecological' farms.
- A documentation centre to collect and document examples of environmentally sound agricultural practices.
- Studies by NGO members of the agro-ecology of their project areas, to increase understanding of the problems and opportunities.
- Training courses for and consultancies to NGO staff on the subject of sustainable agriculture.
- Publication of educational materials on sustainable agriculture, including booklets, posters and slides.
- Publication of a newsletter in the Tamil language.

The participants felt strongly that the network should be semi-formal. A network secretariat should be created to record minutes, handle correspondence, publish a newsletter and maintain a library and documentation centre. Three of the founding members—AME, the Social Forestry Information Project (SFIP) and Kudumbam, a community development-oriented NGO—took responsibility for starting up the network and coordinating its activities for the first 3 years. In January 1991 the Humanistisch Instituut voor Ontwikkelingssamenwerking (HIVOS), a private funding organization based in the Netherlands, agreed to support the network for the period April 1991 to March 1994.

Putting plans into action

In 1991, after funding had been found, staff were appointed for the network secretariat. This now consists of a coordinator, who has extensive rural development experiences and administrative skills, a staff member for the documentation centre, and three agriculturalists covering various areas of specialization. These staff provide support to the network members in the three different 'zones' covered by the network. (From the beginning it had been felt that most interactions between network members should take place within these zones rather than at the state level, so as to reduce members' travelling distances. Apart from the annual general meeting, many network activities are therefore organized within the three zones.)

Organization of the network

Bearing in mind that the staff of the founder organizations all already had full-time jobs related to their own programmes, the structure and administrative tasks of the network were designed sensitively and kept to a minimum. Responsibilities were divided according to time availability, expertise and aptitude. The secretariat is attached to Kudumbam, which is the legal representative of the network and is ultimately responsible for it. Kudumbam, SFIP and AME are responsible for coordinating activities in the three zones. As so-called zonal organizers, they assist in organizing the zonal meetings, the study tours, zonal staff training, data collection for the village agro-ecological study and the documentation of experiences, including traditional knowledge and the experiments of farmers and organizations.

Within the secretariat an editor/documentalist has been appointed. An editorial board of six members meets six times a year to discuss content and policy matters concerning the newsletter.

Achievements

Within the network

Through the agro-ecological studies, understanding of conditions at village level has increased and new NGOs and farmers working towards more sustainable farming systems have been identified. The documentation of environmentally sound techniques has begun, including traditional practices and farmers' experiments with new options. And we are now planning a workshop on journalism for NGO staff and farmers, so as to improve the quality of field reporting, documentation and translation.

The newsletter, *Pasunthalir*, has been launched and is being published regularly. We are continuing our efforts to improve its relevance. Several farmers have recently started a Readers' Forum in the newsletter. Since many farmers in the two states are illiterate, we are thinking of an audio-cassette version of the newsletter.

A start has been made with the library and documentation centre.

As a result of these activities more farmers have started to experiment with new practices on their farms and some NGOs have initiated new environmentally oriented programmes. Areas of experimentation include techniques to improve soil fertility, the use of green manures such as *kolunji*, *danchea* and sword bean *(Canavalia gladiata)*. Most of the network farmers are experimenting with herbal pesticides. There is also a growing interest in traditional land races, the seeds of which are being exchanged among farmers. Experienced members are now being asked to assist others in designing 'ecological' farms and programmes.

A most encouraging feature is the number of farmers and NGOs that have joined the network since its foundation in 1990. The 1993 planning workshop was attended by over 100 members.

Links beyond

The network is establishing relations with other national, regional and global entities working on or promoting LEISA. These relations are very important for the network, allowing access to new technology and experiences, strengthening its conviction and broadening its knowledge base. One of our partners working at a global level is ILEIA, which facilitated the founding of the network by providing seed money to organize the first workshop and the first issue of the newsletter. ILEIA also assisted in establishing contacts with the funding agency. Further support by ILEIA in the areas of information and documentation, and our participation in the international workshop on Networking for Low-external-input and Sustainable Agriculture, held in the Philippines in 1992, have been equally important, broadening our contacts with like-minded networks and organizations and deepening our insight into sustainable agricultural development.

We continue to seek cooperation with organizations in India, especially those working in Tamil Nadu and Pondicherry. These include grass roots organizations as well as government research and extension services. Financial assistance for environmentally sound programmes is now available from national supporting organizations such as the National Wasteland Development Board and the Ministry of Environment. Six NGOs have obtained funds from these organizations to create awareness on biodiversity.

No Start Without Problems

As yet, only a limited number of individuals and organizations in our area have practical experiences in developing LEISA approaches. Many others are interested in such approaches. But they have little idea concerning the scope of LEISA, its implications, how to go about it, and so on. Most of the NGOs interested in working with farmers to promote LEISA lack technical knowledge on agriculture. Training is needed to support them, at least initially. So far two zonal training events have been organized and there are constant requests from members for more.

There is also a need to increase the assistance we provide to NGOs in the analysis of the situation in their areas, thereby helping them to reorient their existing activities in favour of LEISA.

Experience of LEISA in South India has so far provided insufficient evidence of its benefits to convince other farmers to adopt it. The 'ecological' farmers who are founding members of the network are not typical resource-poor farmers, so their experiences are only partially relevant to this group. The most urgent need is to develop technology and approaches suited to the conditions resource-poor farmers have to contend with. The poverty of most farmers, which forces them to sell their labour off the farm, makes this very difficult, since only options which do not need heavy investments in labour or capital will be acceptable. Much more research is needed to adapt existing options and develop new ones.

These challenges place heavy demands on the secretariat and core members of the network, who are also involved in the activities of their own organizations. In short, there is too much work for too few people! Increasing the active participation of other members in the organization and activities of the network is therefore vital. One way of doing this will be to transfer the responsibility for zonal activities from the zonal coordinators to the zonal coordinating committees that are presently evolving.

Finding qualified staff for the network secretariat, with the right attitude and willingness to live and work in the villages in our area, is difficult. Enthusiastic, dedicated individuals can be found, but often they are young and lacking in experience both of environmentally sound agriculture and of networking.

Many network members, especially the farmers, have expressed the need for more frequent get-togethers, apart from the larger annual workshops. They would like more opportunities to discuss the problems that arise during experimentation with LEISA approaches and technologies. In a few areas this has resulted in the emergence of local sub-networks organized by and for farmers only. Farmers who have been with the network from the beginning are assisting their less experienced colleagues with practical information on how to recycle farm waste, on intercropping, sericulture and bio-fertilizers.

Relations with official research and extension agencies have been difficult but are slowly improving. Interest in sustainable agriculture is growing, but there are many differences in attitude, objectives and language—and these are difficult to bridge. Since April 1991 systematic efforts have been made to build up contacts with research institutes. In a few instances this has led to increased interaction between NGO staff, farmers and individual researchers. More recently, Kudumbam and AME have become involved in the local on-farm trials programme of an agricultural research institute.

In conclusion, we feel our network is gradually becoming more effective, and that people from other development circles—such as research, policy and donors—are becoming more enthusiastic about the ideas for which we stand. But without doubt the development of sustainable agriculture in South India still has a long way to go.

Address
Oswald Quintal: Kudumbam 7, Ezhil Nagar, Keeranur 622 502, Pudukkottai Dt, India.
Gandhimathi: AME, P.O. Box 11, Pondicherry 605 001, India.

PART FOUR

Research Networks

Networking in International Agricultural Research

Donald L. Plucknett, Nigel J.H. Smith and Selcuk Ozgediz

Introduction

Networking is a new name for an ancient practice. People have cooperated with one another since the beginning of the human race. But the extent and organizational modes of collaboration have changed markedly since the 1960s, particularly in science. Research networks are proliferating, aided in part by new technologies that facilitate communication among scientists.

Why has this organizational form become so popular in agricultural research? The most obvious advantages are greater efficiency and a cross-fertilization of ideas. A network often uses existing facilities rather than building new ones. It pools research talent rather than adding new staff. Sustainable agriculture depends on inputs from a broad range of disciplines, so collaborative research teams tap scientists from a wide variety of fields.

However, if networks are not carefully planned and managed, there may be more 'net' than 'work'. Research may be constrained if networking procedures consume too much time and money. Too many side-projects may develop in a multidisciplinary network. To avoid these and other pitfalls, it is necessary to understand and evaluate the structure and functioning of research networks and to formulate principles for successful networking.

Drawing on Plucknett et al (1990), this paper is based on information gathered while studying a large number of agricultural networks. It provides insight into the structure of research networks, their problems, and the principles underlying success. It further suggests a conceptual model for evaluating the effectiveness of research networks.

Definition and Typology

A network is an association of independent individuals or institutions with a shared purpose or goal, whose members contribute resources and participate in the two-way exchange of information and/or materials. Two important characteristics of networks are their participatory management and decentralized nature.

To understand networking, it is important to recognize different types of network. As networks usually evolve over time, however, they often fit into more than one category.

They may even be completely transformed. The typology used in this article is based on that used by the Special Program for African Agricultural Research (SPAAR, 1986), classifying networks according to their operational styles. SPAAR's typology distinguishes information exchange networks, scientific consultation networks and collaborative research networks. We add a fourth category: material exchange networks. Table 1 summarizes the characteristics of these four types of network.

Table 1 Some characteristics of networks

Trait	Information exchange	Material exchange	Scientific consultation	Collaborative research
Coordinator	Yes	Yes	Yes	Yes
Publications	Yes	Yes	Yes	Yes
Advisory board	No	Yes	Yes	Yes
Study tours	No	Yes	Yes	Yes
Training	No	Yes	Yes	Yes
Workshops	No	No	Yes	Yes
Common methodology	No	Yes	No	Yes
Joint planning	No	No	No	Yes

Information exchange networks

Despite their name, information exchange networks are often characterized by the one-way dissemination of information—there is little interaction between participants. Participation usually requires no more than being put on a mailing list to receive a newsletter. A two-way flow occasionally develops when individuals on the mailing list provide items to the coordinator to pass on to others.

Information exchange networks usually have a simple structure. They are easy to start, incurring minimal costs. They can be highly specialized and relatively small. It is often difficult to measure impact. Participation is open: anyone can easily subscribe to the mailing list.

Material exchange networks

Material exchange networks may be established to test crop germplasm or finished varieties in different environments, or to coordinate the testing, manufacturing and adaptation of agricultural machinery.

This type of network has a more complex structure. Typically, it has an advisory body, conducts training and organizes monitoring tours. Through international nurseries plant breeders can get a reading on the performance of materials in a wide range of environments. Identical screening methodologies are used to enable results to be compared. The coordinator often plays a strong role, disseminating the trial materials and monitoring the results. Information on results flows back to the nurseries so that they can decide on how to continue the trials. These nurseries are mostly coordinated by the international agricultural research centres because of the complex tasks of collating and disseminating the results.

These networks can play an essential role in the international evaluation of research products. They sometimes develop into collaborative research networks.

Scientific consultation networks

Scientific consultation networks are characterized by two-way communication. Researchers meet at workshops and conferences to exchange ideas and to discuss progress and problems. Their research is locally planned, and projects exist before researchers enter the network. The methodologies used in these projects need not be identical. Training is often an activity of scientific consultation networks.

These networks are easy to start because independent research projects already exist and need no realignment. Sometimes the network is a clearing house for funding requests. Scientific consultation networks often aim to become collaborative research networks, but sometimes the needs of participants are sufficiently met through consultation.

Collaborative research networks

In collaborative research networks, research projects are jointly planned and carried out, using a uniform methodology. However, existing research facilities are used.

Collaborative research networks are coordinated more tightly, and the roles of participants are well defined. The network often has a steering committee, organizes monitoring tours, workshops and meetings, and conducts training courses. The start of a collaborative research network is often prepared by a founding document. Unified research methods lead to the extrapolation of results.

A Conceptual Framework for Studying Network Effectiveness

A common misconception is that networks are not organizations. Networks are a form of organization, albeit a less bureaucratic and hierarchical one than institutions such as universities or corporations. The conceptual framework for studying the effectiveness of networks is therefore built on elements from the literature on the management of organizations. The framework has five major components: network guidance, management of network resources, management of network activities, management skills and teamwork, and a network's links with its external environment (Figure 1).

Figure 1 Conceptual framework for studying the effectiveness of a network

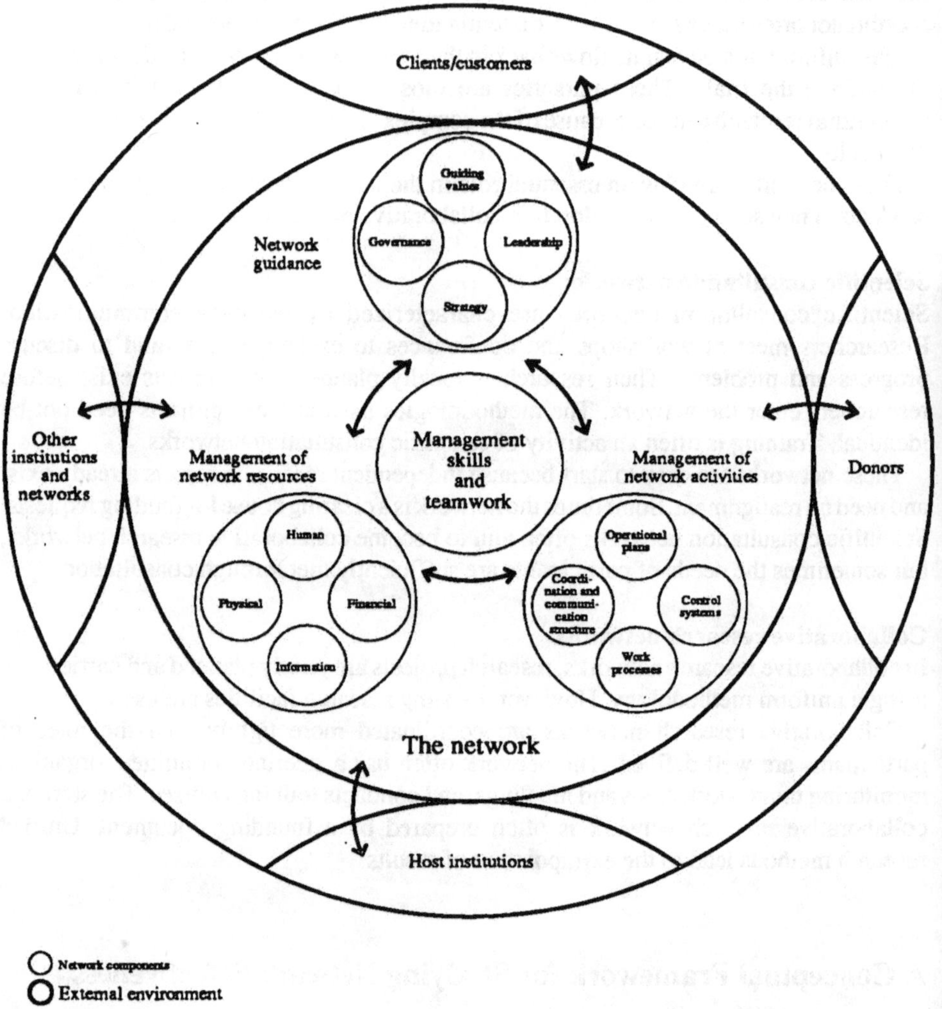

○ Network components
○ External environment

Network guidance
Guiding values are principles and basic ideas shared by members of the network. In agricultural research networks the values of the coordinating body often influence the norms of behaviour and operational style of the network. Values may relate to, for instance, an emphasis on a given region, on publishing in peer-reviewed journals, or on assembling and sharing information.

Leadership is important for networks because of their informal structure. Tasks of the leader include setting a realistic research agenda, finding donors and motivating the weaker participants. The leader or coordinator is often elected. Being the coordinator of a large network is often a full-time job. As a result, this person is frequently employed by a member institution having the necessary office and research facilities.

The governance of a network is usually in the hands of its advisory, steering or working committee. It is the task of this committee to set priorities, oversee the implementation of research and establish good relationships with the network leader.

The strategy of a network is the vision it has of its future. A network strategy identifies the common goal and a path by which to reach this goal. Strategic issues that need to be addressed include how to organize the network, who are its clients, and how to manage resources.

Management of network resources

Human resources are the most important asset of a network. By human resources we mean not only the leader at the core of the network but also the scientists who carry out the network's research. A network can develop its human resources through training programmes to upgrade the skills of the researchers. Monitoring or study tours build expertise and offer opportunities for researchers to discuss ideas and problems. Workshops are another means of cementing the professional bonds between network collaborators. These activities may promote the 'rim effect' of a network, whereby participants collaborate with each other as well as with the network centre, or hub.

Financial resources need special attention as stable funding is extremely important. Start-up funds usually come from one committed donor. Additional funds may subsequently be needed to carry out network activities, but continuing dependence on external funding should be avoided if possible. Producing results is the most effective way of generating support. It is therefore important that network members regularly update their coordinators on the progress of research. Members should also engage in sound financial planning. Network activities need a price tag, so that fund raisers can present a budget to prospective donors. If financial systems are the same for all members, it is easier for the network to monitor its financial situation. Financial credibility is, of course, an important consideration for donors. If a large institution hosts the coordinating body of the network, funding gaps are sometimes temporarily bridged by this institution by means of a loan.

Physical resources, such as laboratories, equipment and vehicles, are widely scattered, although there may be a concentration of more sophisticated facilities for strategic research at the hub of the network. Members need to inform each other of their physical resources and, in some cases, to persuade their employers to allow the network to use them.

Information resources are shared by all network members, who have equal responsibility for acquiring, processing and distributing information. A newsletter is the most common way of disseminating information, including research results.

Management of network activities

Operational plans are the means by which the network's main direction is translated into a set of specific activities. In a successful network, strategic and operational planning are integrated with monitoring and control activities. Plans need to be realistic enough to encourage participants in their work rather than demoralize them. In larger, more complex networks, each network node may have its own plan, but this should fit in with the overall network plan.

Control systems are necessary to enable stakeholders to assess whether network objectives are being achieved, and to allow the network as a whole to respond to its changing environment. Review activities can be undertaken externally, for example by donors, or internally, through monitoring tours, workshops, annual meetings and by the governing body of the network. Each member is usually subject also to local reviews, the results of which can be shared with other members.

Coordination and communication structure

This may vary with each type of network. In information exchange networks, the hub usually collates and processes information. Communication typically flows from the hub to the network members. In other network types, peripheral communication links also exist. In these cases, the hub may be elected and rotate among the members, providing leadership training and sharing the task of administration. Whatever the communication model, it should give all members equal access to decision-making and planning. Good communication enhances decentralization of the network, but because it is voluntary it depends on the motivation of the members.

Work processes include field work, testing, data collection, analysis and similar activities. It is important that the network and its members identify the right processes for meeting specific objectives and implement them effectively.

Management skills and teamwork

These attributes are difficult to quantify yet essential for success. The management skills required by network coordinators include being able to set the goals, plan work, coordinate it, give direction, solve problems, review activities, provide feedback and motivate communication. For good teamwork, a participatory team spirit is necessary. It can be facilitated by an open organizational structure, clear operating procedures and sensitive leadership. Cooperative activities can enhance team spirit.

External environment

Successful networks find ways of manipulating their environment to their benefit. The major players in the network's external environment are as follows:

The clients of a network are those whose interests it is intended to serve. The clients of the networks operated by the international agricultural research centres, for example, are national agricultural research systems. It is important for the network members to be aware of their clients' needs. Regular contact can be established by, for instance, inviting client representatives to network workshops.

Donors are interested in the overall impact of the network. They should be kept informed regularly of its progress.

Host institutions can provide or withold vital forms of support. It is important that host institutions share network objectives and are kept informed of progress. Network members should continually seek support for the network within their own institutions. Policies of the host institution may pose problems for the network member (e.g. on recruitment of staff, financial management). In this case network members should try to adjust their own procedures rather than change their host's policies.

Other institutions and networks should play a role in the network's activities. The strategic plan should explicitly recognize the institutions with which the network expects to collaborate and share responsibilities. Appropriate links can then be established.

Principles for Effective Management

Based on the various facets of our conceptual model, a number of principles for effective network management can be distilled:
- Widely shared values
- Effective governance and policy-making structure
- Effective coordinator/leader
- Clear, well focused strategy
- Decentralized coordination and communication
- Sound, pragmatic work plans
- Effective review mechanisms
- Appropriate work processes
- Good system for recruiting and keeping network staff
- Stable and adequate funding
- Adequate physical resources
- Effective information management and sharing
- Good links with clients
- Productive dialogue with donors
- Healthy relationships with host institutions
- Effective ties with other networks and institutions
- Skilled research managers at the network nodes
- Effective teamwork among participants.

If these principles are adhered to, a simple and flexible network can offer highly effective use of existing research capacity. No major investment has to be made, and there is no painful dismantling once the job is done. If they function well, networks make better use of existing information and help prevent redundancy in research.

How Networks Develop

Each network has its own development path, which can be evolutionary, arising spontaneously from changing needs and priorities, or which can be planned and set in motion from the outset. A network's development is particularly affected by two factors. The first is its articulation, which can be defined as the type, methods and scope of relationships that develop among partners in a network. Articulation becomes more complex over time and therefore needs to be planned. The second factor is differentiation, which refers to the division of responsibilities and related tasks among network members. This also becomes more complex over time. The more explicit articulation and differentiation become, the higher the degree of formalization.

As networks evolve, they may change category (e.g. from a material exchange to a scientific consultation network). This change often requires increased inputs, which need to be anticipated well in advance, particularly where additional financing is involved. In its development from small and new to more complex and mature, a network tends to become increasingly decentralized.

Stages of development

Initiation is the first stage, in which a promoter or instigator plays a catalytic role. Writing a founding document and organizing an initial workshop can be key activities at this stage. The investments in thinking and planning without tangible results are high during the initial stage. It is therefore important that network founders are backed by their institutions. Reasons for starting a network can be to solve international or regional problems, and to build institutional capacity in research.

Early growth is the second phase in the development of a network. Strong leadership is essential to help generate momentum at this stage. The early dissemination of results helps increase the confidence of participants and stakeholders. A growing network is a learning environment for participants. In the early growth stage, the first steps towards decentralization may be taken.

Maturity is characterized by well established articulation and differentiation. It is the most productive stage of a network's development. It is also a potentially dangerous stage, in that procedures and efficiency may no longer be questioned and members may fail to keep abreast of new techniques. Annual meetings are therefore vital to rejuvenate the network. Bureaucracy may be a pitfall at this stage, which may also be characterized by coordinators' relaxing their control, the emergence of sub-networks and the transfer of leadership.

Dissolution of the network takes place when a problem is solved or recognized as unsolvable. The broader a problem, the longer the likely lifespan of a network. Instead of dissolving, however, a network may be transformed. Some questions need to be asked before transformation takes place. Has the network's task been completed? If not, what remains to be done? Are there other opportunities for fruitful cooperation? What changes in strategies and operations are necessary? If neither dissolution nor transformation take place, the network may gradually die a natural death.

Principles for Success

Many principles for successful networking have been identified by various authors. We summarize as follows:

- Problem widely shared. If a problem is recognized as a major one by many parties, donors tend to be interested in funding a network on it, since the potential impact is great.
- Self-interest. Participants should directly benefit from networking activities. If institutions also benefit, they are more likely to support the participants and allow them time and resources for network activities.
- Participants involved in management. Participants should be involved in establishing priorities and planning research. Self-governance should be stimulated from the beginning.
- Clear definition of the problem. Without this the network becomes unmanageable and time is wasted as network members work on subjects that are not relevant.
- Founding document. A baseline study that explores the scope of the problem and identifies key participants is essential for scientific consultation and collaborative research networks. Donors may be willing to donate seed money for a feasibility study that may lead to a network.
- Realistic research agenda. In a well functioning network, each participant is responsible for a piece of the research puzzle that accords with his or her capacities. Unrealistic goals sap motivation.
- Flexibility. Networks need to be flexible to respond to changing research and farming environments. They need self-criticism and periodic corrections.
- New ideas. These are vital to prevent stagnation. New ideas can be generated by linking with other research institutions.
- Regular meetings. These foster the exchange of new ideas and techniques. Meetings are especially important in multidisciplinary networks. However, if meetings are held too frequently, they drain resources needed for research.
- Collaborators contributing resources. If participants contribute their own resources to the research effort, this is a good indicator of their commitment. Collaborators should not be bought: prolonged and heavy subsidies are not a good idea.
- External funding. This is needed for coordination, travel, meetings, etc. Seed money may be needed in the pre-network phase. Especially in developing countries, funding facilitates travelling.
- Training. This is necessary to upgrade capabilities and bridge gaps in expertise between partners at the start of a network, or between new and old members.

- Stable membership. This promotes continuity and a collegial atmosphere. Valuable time is lost when a constant stream of new members have to be informed of networking procedures.
- Strong leadership. Dissatisfaction is less likely to occur when a leader is elected rather than imposed. If leadership changes too frequently, the network's cohesion suffers. However, when a network is mature it is easier to change leaders without disrupting activities.

Problems and Remedies

Networking problems can be classified into three groups.

Research quality is one field where problems may occur. The network may either have difficulties in acquiring and processing sufficient data, or the quality of the data may be inadequate. Causes could be communication difficulties, poor information management, poor feedback (leading to lack of motivation), lack of trained scientists, shortages of functioning equipment, communication difficulties between biological and social scientists, and so on. In the case of varietal trials, feedback problems can be solved by prescreening the trials and sending out preliminary reports after early feedback has been received. To allow results to be compared, it is important that the different locations should be clearly described. One problem that needs to be solved is that even descriptive methods vary.

Personnel problems may occur when there is a rapid turnover of participants. Delay in the appointment of a successor creates a lack of continuity. The successor may not be as well motivated. This is especially dangerous in the early stages of a network. To remedy personnel problems, the network may need to create better career prospects and more attractive training programmes. Language may appear a problem in some international networks, but where there is an atmosphere of mutual respect and methodology and terminology are clearly defined, language barriers can usually be overcome.

Institutional and bureaucratic problems often encountered are poor planning, unclear agreements with governments, bad timing and coordination of activities, poor funding arrangements (shortage or uneven flow of funds) and withdrawal of important donors. A shortage of funding can be remedied by making sure that results are delivered quickly. This can be achieved through proper planning, combined with a clear goal and a realistic research agenda. If important donors suddenly withdraw, successful networks are able to tap funds to bridge gaps. These could be in the form of loans. International centres are well suited to acting as the hub of a network, because they often have greater financial credibility and so can overdraw their accounts more easily. Financial risks can also be remedied by pooling funds from several donors, although this may create a new level of bureaucracy. Bad accounting may lead to financial problems. It can be caused by differences in the accounting methods of network members. Another reason may be that members do not produce reports on time, leading to a drawback of funds.

Other bureaucratic problems may be caused by governments, which may have different research priorities to those of the network. On the other hand, networks can dominate or manipulate national programmes. This happens especially when the need for a network is expressed by a donor rather than by network participants.

Sometimes there is a gap between a network's potential and its actual performance. Alternatively, successful research results attract so many new projects and participants that the network becomes unmanageable. Remedies for problems of this kind are dissolution, splitting up, transforming or creating sub-networks.

Conclusions

Networking offers many benefits. It can link national research with regional and international efforts. It can increase efficiency—more research can be carried out at lower cost. It can improve methodologies. Networks enhance interaction among scientists. They upgrade the skills of the weakest links. When well managed they can create trust, confidence and collegial bonds. Most important, they can bring new insights and fresh lines of inquiry, leading to the generation of new technologies. Network members actively participate in the creation of technology instead of being its passive recipients. In this case the technology is more likely to be relevant to national needs. Finally, networking can strengthen the research capacity of participating institutions. Existing knowledge and data are more efficiently used.

However, too many networks may lead to the duplication of research and too much competition for too little funding. In addition, weak national programmes may become overloaded. Donors are partly to blame for the current overlap between different networks, as they sometimes have special agendas in funding networks.

References

SPAAR. 1986. African agricultural research networks: Summary papers and tables. Meeting of the Technical Working Group on Networking of the Special Program for African Agricultural Research (SPAAR), Brussels, Belgium, July 7-8.

Plucknett, D.L., Smith, N.J.H. and Ozgediz, S. 1990. *Networking in international agricultural research.* Cornell University Press, Ithaca and London.

Address

Donald Plucknett, CGIAR, World Bank, Room N 5055, 1818 H Street, Washington D.C. 20433, USA.

Animal Traction Networks in Africa: Lessons and Implications

Paul Starkey

Introduction

In many parts of the world, animal traction is an appropriate, affordable and sustainable technology, requiring few external inputs. Work animals can be used to reduce drudgery and intensify agricultural production, so raising living standards throughout rural communities, benefiting men and women, young and old. Cattle, buffaloes, donkeys, mules, horses, camels and other working animals can provide smallholder farmers with vital power for crop cultivation and transport. Draught animals can also be used for other activities, including water-raising, milling, logging, land-levelling and road construction.

In North Africa and the Nile valley, there has been a very long history of animal traction. A large number of draught animals, including oxen, cows, bulls, donkeys, mules, horses, buffaloes and camels, have been used for soil tillage and transport. There has also been a long tradition of using work animals in parts of the Horn of Africa. In Ethiopia, which has the highest population of draught animals in Africa, traditional cropping systems almost invariably involve the use of the wooden *maresha*, which is an ard plough pulled by a pair of work oxen. Pack donkeys and mules are also widely used in Ethiopia. Elsewhere in sub-Saharan Africa, animals have long been employed for transport by certain pastoralists and traders, but animal-drawn implements have not been widely used in traditional farming systems.

Animal traction for tillage and wheeled transport was introduced into sub-Saharan Africa during the colonial period. Indeed, in most African countries it was pioneered during the lifetime of the present elders. The technology, usually involving pairs of work oxen and imported metal implements, spread slowly during the first half of this century. There was great variation in adoption rates, with fastest adoption in areas with relatively developed crop marketing systems, particularly for cotton and groundnuts.

During the 1960s and early 1970s animal traction received relatively little attention from newly independent governments. This was a period when many people thought that the rapid tractorization recently seen in Europe and North America should also take place in African countries. Animal traction had dropped out of the curriculum in Europe, and it was also often omitted in sub-Saharan Africa. A generation of agricultural students graduated with little or no formal training relating to animal traction. These agriculturalists were often rapidly promoted within ministries and research organizations and became responsible for planning and implementing agricultural projects and programmes.

By the late 1970s, higher oil prices, foreign exchange shortages and numerous failed tractor schemes suggested that rapid motorization was not, after all, practicable. In contrast, animal traction began to seem appropriate, affordable and sustainable. Research suggested that animal traction could reduce drudgery and increase crop production (mainly through area expansion), and that many social and economic benefits could arise from the introduction of animal-drawn carts. Animal traction started to be seen in many countries as a serious, but neglected, development option.

With the inflow of donor funds that followed the well publicized Sahelian droughts, many donor-assisted projects were established in Africa to introduce (or re-introduce) and/or conduct research on animal traction technologies. These projects tended to work in isolation, unaware of each other. Many were oriented to solving technological constraints, and ignored social and economic factors. Several experienced serious problems because those implementing the project did not really understand all the technical, social and economic implications of using animal traction (Sargent et al., 1981; Munzinger, 1982; Starkey, 1986).

International Information Exchange

In 1982, the Food and Agriculture Organization (FAO) of the United Nations convened an expert consultation on animal traction. This concluded that improved information exchange on draught power was extremely important (FAO, 1982; 1984). As a follow-up, FAO, in conjunction with the International Livestock Centre for Africa (ILCA), then organized a series of missions to 12 African countries in 1983, 1984 and 1985 to investigate the possibilities of establishing an animal traction network in Africa (Imboden et al, 1983; Starkey and Goe, 1984; 1985).

The missions found that there was very little information exchange taking place between animal traction programmes within countries, let alone between countries. There were far too many cases of projects, often only a short distance from each other, 'reinventing the wheel' (or redesigning an implement) in almost total isolation. The missions concluded that a network was not only extremely desirable; it was also highly feasible. There existed strong support for the idea both at project/institutional level and in the national ministries. It was suggested that it might be most practicable if a network were to be launched in West Africa, to be followed quickly by complementary initiatives in Southern and Eastern Africa (Starkey and Goe, 1984; 1985).

The West Africa Network

Although the FAO/ILCA proposals had stimulated interest in the creation of a network, for various organizational and institutional reasons there was no immediate follow-up. Rather, the practical initiative that led to the creation of the network was a small workshop organized in March 1985 by the Farming Systems Support Project (FSSP) of the University of Florida. FSSP had identified animal traction as one area in which a farming systems perspective was desirable, and one means by which crop and livestock production (which were often seen as separate in West Africa) could become more integrated. The workshop was hosted by an animal traction project funded by the United States Agency for International Development (USAID) in Togo.

For several anglophone and francophone countries this 1985 'networkshop' was probably the first time that people in West Africa had come together specifically to discuss animal traction technology and to review it from a farming systems perspective. The 30 participants highlighted technical, economic and infrastructural constraints and debated the preconditions for the successful development of animal traction (Poats et al, 1986). They regarded the workshop as extremely useful, and resolved to hold a follow-up workshop which would allow further in-depth analysis of the issues and enable more countries in West Africa to exchange information.

A steering committee was elected, comprising representatives from the animal traction programmes of five West African countries, a representative of the main resource organization (University of Florida) and a facilitating technical adviser. Thus the West African Animal Traction Network (WAATN) was born. The committee met later in 1985, in The Gambia, and invited Sierra Leone to host the next 'networkshop'. It also recommended several activities to improve information exchange between countries and with other networks. For example, two committee members took part in a study tour of animal traction organizations in Nepal and Indonesia, and circulated a report on the implications for programmes in West Africa (Starkey and Apetofia, 1986).

Workshops and workshop methodology

Among the most prominent activities of the West African network have been the major workshops, organized every 2 years. In 1986 a workshop on Animal Power in Farming Systems was held in Sierra Leone. This was attended by 73 people from 20 countries, with 34 papers written by 51 people active in animal traction being circulated and published in the proceedings (Starkey and Ndiamé, 1988).

This was followed by a workshop on Animal Traction for Agricultural Development, held in Senegal in 1988, which was attended by 78 people from 24 countries. Some 60 papers were published in the proceedings (Starkey and Faye, 1990).

In 1990, a third workshop, on Research for Development of Animal Traction, was held in Nigeria, and was attended by 93 people from 19 countries. Circulated at this workshop were 52 papers prepared by 75 people working in animal traction (Starkey, 1990a). The proceedings are being edited for publication in association with ILCA.

Thus, to date, network workshops have been attended by over 200 people. Furthermore, the workshops have directly stimulated the preparation and publication of over 140 papers covering a wide variety of issues and experiences concerning animal traction in different farming systems and related research, development and policy implications.

The workshops have proved extremely popular, and participants have considered them interesting and professionally valuable. The evaluations conducted at the end of each have allowed the organizers to learn which aspects have been most appreciated. All three workshops have used the same general approach and methodology, with variations based on local conditions and on the feedback from the previous evaluation.

The workshops have been well publicized in advance, with an open invitation to all those working in the field of animal traction, in West Africa and elsewhere. This open approach has encouraged a broad range of people to attend, and has been unlike the 'closed' international workshops more commonly organized in Africa, where attendance is only by specific invitation to individuals or officially nominated representatives.

Although the invitation has been open, certain conditions have had to be met, including the submission of a suitable paper. Furthermore, when too many people from the same country have applied to attend, selections have had to be made based on quality of papers and the need for a suitable balance of different organizations and disciplines.

As a result of the open invitations, the workshops have been thoroughly multidisciplinary, bringing together agricultural engineers, economists, animal scientists, agronomists, sociologists and other professions. Furthermore, the participants have come from different professional fields, with researchers, extensionists, administrators, producers and donor representatives all closely interacting.

Although participants have received copies of all the papers prepared, they have not spent much time sitting through long sessions of paper presentations (which people tend to find tedious). Rather, there have been a few selected key papers, designed to stimulate discussion. Informal discussion has also been stimulated by 'networking announcements', in which people have had an opportunity to briefly summarize their work and interests, and the topics on which they would like to exchange information during the week. Sometimes these have led to special evening sessions for those with particular interests, and these have led to subsequent collaboration. For example, at the 1990 workshop in Nigeria, participants from Eastern and Southern Africa met in one special session to discuss the formation of an animal traction network for that region.

Without doubt, the most popular elements of each workshop have been the field visits. People who have been to conferences where the field visits have involved large groups slowly straggling around research sites may be surprised at this. The network field visits appear to have been popular because they have consisted of small groups of five to eight people from different countries, who have gone to villages to watch work animals in use and to listen and talk to farmers. Such in-depth discussion with farmers is accepted as an integral part of the farming systems approach, but has often been a new experience for workshop participants. They have often felt free to ask farmers questions they would never have dared to ask in their own countries, for fear that their juniors would laugh at them. The small groups have also visited village blacksmiths. Some groups, returning from the villages, have briefly visited project sites, research stations and implement producers.

In the day following the field visits, the small groups have sat down to discuss in detail their observations and findings, and to review specific themes highlighted in the key papers. The groups have then reported back to all the other participants, in preparation for open discussion on the issues raised. The small group discussions have proved almost as popular as the field visits.

The workshops have also provided an opportunity for a network business meeting, to discuss plans for the network, and elect a new steering committee to supervise the forthcoming programme.

Network publications

A further important element of the workshops has been the publication of the proceedings in an attractive format. Copies have been made available free-of-charge to people working in Africa. As there are no specific animal traction journals, people have tended to publish their experiences in the periodicals of their particular discipline, including journals of anthropology, agricultural engineering, economics and animal science. Unfortunately, even in countries blessed with well stocked libraries, these journals are seldom read by their colleagues of different disciplinary backgrounds who are also working with animal traction. In Africa, such specialized professional journals are only rarely available to people actually engaged in animal traction research and development. Publishing workshop papers in a single volume has thus provided useful and easily accessible information for those working in this field. Furthermore, non-participants, seeing such proceedings, have been encouraged to put their own experiences in writing for subsequent workshops. To date three proceedings have been published (Poats et al, 1986; Starkey and Ndiamé, 1988; Starkey and Faye, 1990) and one is currently being prepared.

The German Appropriate Technology Exchange (GATE) has also published a series of other animal traction resource books based largely on the networking experience and approach, including the *Animal Traction Directory: Africa* (Starkey, 1988a). ILCA has published an animal traction bibliographic database, made possible through the same networking approach (Starkey et al, 1991). These publications have been made available free-of-charge to network members in Africa.

Other network activities

Between the main workshops described above, the network steering committee has met once or twice a year, as far as possible each time in a different country. As these meetings have been combined with field visits, they have been, in effect, small group study tours, with mutually beneficial interactions between the committee members and representatives of the host country.

Other activities have been carried out collaboratively by two or more country programmes or by the members of special interest groups. For example, in 1989 ILCA hosted a planning workshop for WAATN members specifically interested in collaborative research.

This was held to develop consistent research protocols for implementation in West Africa. The end result, once funding is obtained, should be a well coordinated collaborative research programme (ILCA, 1990). The various experiments and projects will be coordinated by a full-time animal traction networking research scientist, based at ILCA's office in Kaduna, Nigeria. This research scientist also represents ILCA on the network steering committee.

As with all members of the network, the research coordinator is free to communicate directly with other members—a further illustration of the open and informal nature of the network. Similarly, visits and collaboration have been arranged between (for example) Sierra Leone and Togo, Senegal and the Gambia, Guinea and Mali. Collaborative research has also been arranged between (for example) the French Centre d'Etudes et d'Expérimentation du Machinisme Agricole Tropical (CEEMAT) and national research groups in Senegal and Burkina Faso, and between GATE and Senegal. These activities have been arranged directly between members of the network, and may or may not have been stimulated by contacts made during network workshops. They are considered to come under the network umbrella in that they involve collaboration between members, with the information produced likely to be reported in subsequent network workshops and also diffused informally through other networking contacts.

Although there is no official network newsletter, one country, Togo, produces a national animal traction newsletter, Force Animale, which it circulates to several other network members (PROPTA, 1991). Other document exchange continues on an individual-to-individual or organization-to-organization basis. Thus documents produced in Mali, Sierra Leone, Togo and Senegal (for example) are now quite commonly found in other countries in the region. This was not the case 6 years ago, when the network was launched.

The Eastern and Southern Africa Network

In 1987, the Southern African Centre for Cooperation in Agricultural Research (SACCAR) organized a regional animal traction workshop in Maputo, Mozambique. At this it was resolved that a regional information-sharing network should be established under the auspices of SACCAR (Namponya, 1988). For institutional and organizational reasons there was no immediate follow-up to this, but several individuals from Eastern and Southern Africa participated in animal traction workshops organized in 1988 (Senegal), 1989 (Indonesia) and 1990 (Scotland and Nigeria). On each occasion, the participants from the region affirmed that they should form their own animal traction network.

As a direct result of the 1990 workshops, two separate networking initiatives in Eastern and Southern Africa were started. For a few months they coexisted as parallel schemes, but they came together in 1991. One was initiated by staff of Christian Mission Aid (CMA), a non-government organization (NGO) based in Kenya. The other involved animal traction specialists based in Zambia and Zimbabwe.

A valuable opportunity to launch the animal traction network for Eastern and Southern Africa came in November 1990. The setting was a regional course on planning integrated animal draught programmes, held at the Agricultural Engineering Training Centre (AETC) of the Institute of Agricultural Engineering in Harare, Zimbabwe. The course was arranged by AGROTEC (Programme on Agricultural Operations Technology for Smallholders in East and Southern Africa), a regional project of the United Nations Development Programme (UNDP), funded by the Swedish International Development Agency (SIDA). During the course, there had been much discussion about networking, and the experience of the WAATN had been presented. The course participants therefore selected six people from different countries to form a committee to discuss organizational details and prepare an action plan for the new network. Representatives of AGROTEC, the Gesellschaft für Technische Zusammenarbeit (GTZ) and a consultant resource person were invited to join the committee. The decisions of this committee to launch the network and organize a major workshop were endorsed by the final plenary session of the AGROTEC course (Kalisky, 1990).

The provisional steering committee of the new Animal Traction Network for Eastern and Southern Africa (ATNESA) met again in Zambia in April 1991 to discuss network organization and to plan the first major open workshop. The chairman of the committee had prepared a paper on possible ways of coordinating the network, and another member had prepared draft statutes, based on those of WAATN. The committee decided to adopt an informal system of network organization, based on national networks linked through a regional network steering committee.

This organizational strategy was due to be discussed at the first open workshop of ATNESA to be held in Zambia in January 1992. The theme of this first workshop was to be 'Improving animal traction technology', and about 100 participants from 20 countries were expected to attend. The workshop was expected to follow the pattern established by the West African network, with the emphasis on field visits and small group discussions. All participants were expected to prepare papers, which would be published in the proceedings. Invited key papers were being prepared collaboratively, with authors in two or more countries (or resource organizations) combining their experiences prior to the workshop. The workshop was also offered as a means whereby members with special interests could meet to discuss their areas of concern. Among the special interest groups likely to meet were those concerned with gender issues, farming systems research, animal-drawn transport, implement supply and distribution, the preparation of animal traction manuals and the use of donkeys.

Network Typology

Both WAATN and ATNESA have evolved as semi-formal, regional networks that focus on animal traction issues. The networks are open to all persons and organizations sympathetic to their aims and objectives, whatever their discipline and whatever their role in animal traction development. Thus researchers are members, but the networks are not limited to research interests. Planners, extension workers, veterinarians and implement manufacturers are all active members. The networks are open to government services, NGOs, cooperatives and private companies. In principle farmers or farmers' groups could also be involved, but in practice farmers' interests are represented by those individuals and organizations working with, or for, farmers (directly or indirectly; perfectly or imperfectly).

Both networks have concentrated on information exchange, which primarily benefits network members. Nevertheless, both networks have had some external orientation and have included the general promotion of animal traction within their network objectives.

Some Lessons

One clear lesson that emerges is that network activities are more important than formal structures. Despite its lack of a formal secretariat, WAATN has been active for about 6 years, and it has much to show for its work. While network members agree that a strong, active coordination unit would be highly desirable (and is in the process of establishment), the absence of this need not prevent a network from flourishing, provided the members are themselves active.

Another important lesson is that keeping the network open and informal makes communication channels more reliable and efficient. Naturally, national and institutional protocols have been respected, but within such limitations network members have been encouraged to correspond directly with their colleagues in other countries. Such direct contact, combined with copying relevant correspondence to interested parties, has proved very effective.

Some network members have argued that all communications should be channelled through a central secretariat and/or through focal points within each country or organization. However, practical experience has shown that both individuals and institutions can suddenly change from being facilitators to being bottlenecks. Whatever the good intentions of nominated representatives, they can, with little or no warning, be promoted to a different post, sent on study leave or incapacitated by illness or an accident. Within national and international institutions, managements can change or shift priorities, work loads can suddenly increase, key staff may leave and budgets can suddenly be cut. In such circumstances, network correspondence, having once been a priority, can be neglected. This may not be too critical if just one individual or organization is involved, but if network members rely on that focal point to disseminate information, several network members may be deprived.

Perhaps the strongest feature of the two networks is that they are informal African organizations. They did not arise from project documents of donors, nor were they created by any one institution. They have grown up through strong member interest and close collaboration with a variety of donor organizations. The networks have received support from several donors and international institutions, but they are not dependent on, nor controlled by, any single one of these. Such arrangements should allow the networks to survive the vicissitudes of policy and funding strategy of specific organizations. Multi-donor support also reduces the risk of any one funding agency using its financial muscle to impose its policies and priorities on the network.

To date, the networks have had no financial resources of their own. They have found that sufficient funds can generally be generated for network activities that have clear objectives, such as workshops and study tours. Often, when the networks have taken an initiative and organized an activity, the costs of travel and participation have been largely met by projects within the member countries.

At all the workshops held so far, the majority of participants have been funded from sources (often donor-assisted projects) within their own countries, and not from the central workshop budget. This makes workshop organization easier and cheaper, and emphasizes the user-supported nature of the networks. When applications for sponsorship have been received by the committee organizing a workshop, it has often been possible to put the applicant in touch with a sponsoring organization within his/her country, thereby initiating useful and beneficial contacts.

In the past, the networks have received support from the FSSP, GTZ, GATE, ILCA, AGROTEC, the International Development Research Centre (IDRC), the Technical Centre for Agriculture and Rural Cooperation (CTA), the International Institute of Tropical Agriculture (IITA), Environment and Development in the Third World (ENDA), the Netherlands Directorate General for International Cooperation (DGIS) and several national organizations and projects within Africa.

Some Problems

The networks have also had problems. Postal services and telecommunications between African countries can be slow and unreliable. Indeed, because intra-African communications are often less well developed than Europe-Africa links, the fact that the Technical Advisor has had an office in Europe has frequently proved valuable in facilitating liaison and information dissemination.

Air schedules and connections within Africa are such that committee members and workshop participants can seldom all arrive and depart on the same day. Two or even three days may be needed for air travel between some countries. Thus attendance at a 3-day meeting may require people to sacrifice a week from their work. Difficult air schedules can significantly increase meeting costs, as provision has to be made for additional per diem payments.

During workshops and meetings network members can devote themselves fully to network activities. In the enthusiasm of a workshop, participants find it easy to offer to take on responsibilities. However, good intentions can remain just that when members return to their families and to the day-to-day realities of their own demanding jobs. Furthermore, not all members involve themselves in national-level networking activities. A combination of national and international networking is essential to maximize and multiply the benefits of the networks.

One possible danger with any network is the tendency for it to become inbred—familiarity tends to diminish the intensity of communication and the cut-and-thrust of criticism on technical matters when colleagues meet each other frequently. This pitfall has largely been avoided by attracting many new people to each workshop—a policy that has necessitated large workshops. If workshops were smaller in size, perhaps restricted to just one or two participants per country, the same individuals would tend to be involved each time.

The steering committee of WAATN has been fairly constant since its inception. This has given valuable continuity and stability, but the limited turnover of committee members has restricted opportunities for fresh vision and new dynamism. To avoid this, the provisional ATNESA committee decided to recommend that no ATNESA committee member should serve for more than two terms.

It has proved very difficult to bring together all members of the steering committee at the same time. For example, between October 1990 and December 1991, there were three meetings of the WAATN steering committee, but at none of them did the committee members feel that they had enough members present to make binding decisions on the future organization of the network. At each meeting, one or more people crucial to the topic were unavoidably absent, due to conflicting activities, communication difficulties, travel problems, illness, political upheavals or other unforeseen circumstances.

Some resource organizations have tended to be rather fickle, sometimes supporting the networks strongly, sometimes appearing cool. Such inconsistencies have occasionally been brought about by changing institutional policies or by different budgetary situations within the resource organization. At other times they have been attributable to the whims of individuals. Whatever the stated position of a resource organization, practical support for the networks depends largely on the enthusiasm (or otherwise) of one or more key individuals within it. Whether or not an activity is supported depends to a large extent on the prevailing work load, mood or self-interest of the contact person.

Network organization
The networks have been run on a voluntary basis, with no full-time staff. When activities such as workshops have had to be organized, or papers written and edited, the host institutions have given permission for their staff to spend time on these jobs. However, they have not reduced their other work loads, with the result that the individuals assigned to the extra duties have often been quite stressed. The work of the Technical Advisor has

also been largely on a voluntary basis, supplemented by some short-term consultancy assignments, provided by various sponsoring organizations for specific organizational or editorial tasks.

To date, the networks have had no central budget or account. The day-to-day costs of networking have been met by individuals or their parent organizations. Specific activities, such as network committee meetings, study tours and workshops, have been funded by one or more donors, and the organizers and participants have generally been able to claim relevant expenditures from the workshop account or from one of the sponsoring organizations. Most activities have been organized on trust, with outlays made long before the refund. This has placed considerable strain on some individuals and organizations.

Institutionalization

While the current institutional arrangements have worked, they have been far from ideal and various proposals have been made to institutionalize the networks. The steering committee of WAATN has twice prepared ambitious project documents. These have had budgets large enough to hire and house a full-time network coordinator (assumed to be a West African with an international salary), equip a secretariat and provide operating expenses. Donors have rejected these as being too expensive. One point made in jest— but many a true word is spoken in jest—was that the animal traction network has operated effectively for 5 years without a large budget, so this makes it difficult to justify a major financial provision.

One donor has offered to provide funds that would allow committee members to take time off their main jobs and work for a few months on specific network activities. This creative proposal has been received with mixed feelings, and it has been perceived as a second-best to full-time coordination.

One resource organization, ILCA, offered in 1988 to coordinate an animal traction research network from its headquarters in Addis Ababa. This was to be a formal research network, drawing on the informal WAATN. A steering committee would have had overall responsibilities for the network, but day-to-day coordination would have been undertaken (and paid for) by ILCA (Goe, 1988). This offer was put to the general assembly of WAATN, but was politely declined, mainly because of fears of losing control of the network to one member institution with its own goals and priorities. People were under the impression that international research centres had, in the past, used networks to promote their own interests rather than those of the network members. Furthermore, international research centres had mandates limited to research, whereas WAATN wished to ensure that its network continued to also serve the needs of those more concerned with development, extension and implement production.

Negotiations were entered into to associate WAATN with the West African Farming Research Network (WAFSRN). This had moved out of an international research centre (IITA), and had established an independent secretariat in Ougadougou under the umbrella of the Semi-Arid Food Grain Research and Development (SAFGRAD) programme of the Organization of African Unity. Draft protocols of understanding were drawn up with both WAFSRN and SAFGRAD. It was envisaged that WAATN would continue to operate as a fully independent network, under SAFGRAD, sharing offices with WAFSRN.

The steering committee of WAFSRN subsequently decided that, if the animal traction network wished to share WAFSRN facilities, it should become a sub-network of WAFRSN. The WAATN committee did not want to become a sub-committee, and was worried lest its network be 'swallowed up' by WAFRSN and lose its identity built up over the years.

ILCA subsequently offered to host the animal traction network secretariat at its office in Kaduna, Nigeria. Although the secretariat would be located at ILCA, network independence was promised and the name could be retained; there was no need to downgrade to a sub-network.

A final decision still has to be taken, but after more than 2 years of negotiation WAATN still does not have a full-time secretariat or network budget. Although discussions have always been harmonious, there remain unresolved conflicts of interests: the network will almost certainly have to lose some of its independence if it is to have a full-time coordinator (or sub-coordinator).

Ascertaining the benefits

While all those associated with the networks can point to the advantages to individuals and to programmes of improved knowledge and understanding, it is extremely difficult to actually measure the benefits.

If one looks back to the years of work wasted in the past on unsuitable technologies in Africa (such as wheeled toolcarriers, which were 'perfected yet rejected'), one can see the great potential for savings through networking (Starkey, 1988b). For example, one project in West Africa spent about US$ 2 million attempting to introduce Asian water buffaloes as work animals in part of the Sahelian zone of West Africa (Starkey, 1990b). This animal traction project (which was planned before the start of the network) lacked a farming systems orientation. Nor did it benefit from networking interactions with colleagues familiar with other attempts to introduce exotic work animals into sub-Saharan Africa. In retrospect, it seems likely that the money allocated to this project would have been better used had those responsible for planning and implementing it been exposed to the experiences and perspectives of network members.

Other, more recent projects may well have been made more relevant and productive because those designing them have been able to learn from the networks. It is impossible to know how many programmes and projects have benefited, but some clear examples of network influence can be documented.

An animal traction project in Guinea serves to illustrate the genuine yet elusive nature of the benefits. This project has not yet participated in any formal network activity such as a workshop, but its leaders recently made use of some of the network publications to learn of, and then to contact, colleagues working in Mali, Senegal and Sierra Leone. This led to a 3-week training visit in Mali, the testing of Senegalese and Sierra Leonean implements in Guinea, detailed discussions on technical, economic and organizational issues, and the acquisition by the project of documents on a wider range of topics. Moreover, each contact led on to others: for example, the professionals in Mali were able to discuss the experiences of their colleagues in Togo, whom they had met at a networkshop. This project acknowledges that its contacts were made as an indirect result of the network's activities and publications. Such information exchange would have been virtually impossible a mere 5 years previously, simply because people were almost completely unaware of each others' activities. As a result, the project implemented some tried and tested approaches and so achieved in 2 years what in more normal circumstances might well have taken it 3 to 4 years (Starkey, 1991). Significant savings in time and costs were achieved through networking.

While all involved in the Guinea project believe they saved time and money, it would be difficult to prove a cause-and-effect relationship, since so many other factors were involved. It would also be difficult to measure the specific economic benefits, as the time saved could not be quantified without a control. Similarly, it is impossible to quantify the benefits of the numerous similar exchanges that are now taking place within (and beyond) the region. While the value of networking is reflected in the gains in knowledge and project design, it will remain difficult to estimate the total benefits to the region.

Conclusions

There has been a huge increase in the exchange of information on animal traction in West Africa in recent years, much of it due, directly or indirectly, to the activities of WAATN. There are also increasing numbers of examples of collaboration between programmes, notably in areas of research, training and implement testing. ATNESA has started to achieve similar benefits in Eastern and Southern Africa.

While the large general workshops are likely to remain popular for some time, particularly among those for whom they are a completely new experience, it is probable that the networks will place increasing emphasis on events for special interest groups. For example, intensive seminars may be held for researchers working on similar topics (e.g. the use of draught cows), or for development projects involved in similar work (e.g. the use of animal traction for rice production), or for the many implement manufacturers in the two regions. Such activities may be arranged by a network secretariat, or may continue to be organized by one or more network members.

It is likely that the combination of member enthusiasm, open membership, flexible communication channels and multi-donor support will ensure the continuing effectiveness of both networks. For the foreseeable future these networks should therefore be in a position to promote the development of animal traction as a low-cost technology for increasing and sustaining the productivity of smallholder farming systems in Africa.

References

FAO 1982. Report of the FAO Expert Consultation on the Appropriate Use of Animal Energy in Agriculture in Africa and Asia, held in Rome, 5-19 November 1982. FAO, Rome, Italy.

FAO 1984. Animal energy in agriculture in Africa and Asia. Animal Production and Health Paper No. 42, FAO, Rome, Italy.

Goe, M. R. 1988. Animal traction research network: Proposed implementation and operation. Paper prepared for discussion at the West African Animal Traction Network Workshop, held 7-12 July 1988, Saly, Senegal. Mimeo, International Livestock Centre for Africa (ILCA), Addis Ababa, Ethiopia.

ILCA 1990. Annual report 1989. International Livestock Centre for Africa (ILCA), Addis Ababa, Ethiopia.

Imboden, R., Starkey, P. H. and Goe, M. R. 1983. Report of the preparatory consultation mission for the establishment of a TCDC network for research, training and development of draught animal power in Africa. AGA Consultancy Report, FAO, Rome.

Kalisky, J. (ed) 1990. Proceedings of a regional course on planning an integrated animal draught programme, held in Harare, Zimbabwe, 5-13 November 1990. Bulletin No. 2, Agricultural Operations Technology for Smallholders in East and Southern Africa (AGROTEC), Harare, Zimbabwe.

Munzinger, P. (ed) 1982. Animal traction in Africa. GTZ, Eschborn, Germany.

Namponya, C. R. (ed) 1988. Animal traction and agricultural mechanization research in SADCC member countries. Proceedings of a workshop held August 1987, Maputo Mozambique. SACCAR Workshop Series 7, Southern African Centre for Cooperation in Agricultural Research (SACCAR), Gaborone, Botswana.

Poats, S. V., Lichte, J., Oxley, J., Russo, S.L. and Starkey, P. H. 1986. Animal traction in a farming systems perspective. Report of a networkshop held at Kara, Togo, 3-8 March 1985. Network report No. 1, Farming Systems Support Project (FSSP) University of Florida, Gainesville, USA.

PROPTA 1991. Force animale. Bulletin technique trimestriel du projet pour la promotion de la traction animale (PROPTA), Atakpamé, Togo.

Sargent, M. W., Lichte, J. A., Matlon, P. J. and Bloom, R. 1981. An assessment of animal traction in francophone West Africa. Working Paper 34, Department of Agricultural Economics, Michigan State University, East Lansing, Michigan, USA.

Starkey, P. H. 1986. Draught animal power in Africa: Priorities for development, research and liaison. Networking Paper 14, Farming Systems Support Project (FSSP), University of Florida, Gainesville, USA.

Starkey, P. H. 1988a. Animal traction directory: Africa. Vieweg (for German Appropriate Technology Exchange), GTZ, Eschborn, Germany.

Starkey, P. H. 1988b. Perfected yet rejected: Animal-drawn wheeled toolcarriers. Vieweg (for German Appropriate Technology Exchange), GTZ, Eschborn, Germany.

Starkey, P. H. 1990a. Research for development of animal traction: A report of the fourth workshop of the West African Animal Traction Network, held 9-13 July 1990, Kano Nigeria. Animal Traction Development, Reading, UK.

Starkey, P. H. 1990b. Water buffalo technology in northern Senegal. Report prepared for USAID-Dakar (contract 685-0281-000-0199-00) and Projet Buffle, Saint Louis, Senegal. Tropical Research and Development Inc., Gainesville, Florida, USA.

Starkey, P. H. 1991. The revival of animal traction in Kindia Region of Guinea Conakry (Relance de la traction bovine dans la région de Kindia, Guinée Conakry). Project evaluation report ONG/78/89/B, Guinea Conakry. Commission of the European Communities, Brussels, Belgium.

Starkey, P. H. and Apetofia, K. 1986. Integrated livestock systems in Nepal and Indonesia: Implications for animal traction programs in West Africa. Farming Systems Support Project (FSSP), University of Florida, Gainesville, USA.

Starkey, P. H. and Faye, A. (eds) 1990. Animal traction for agricultural development. Proceedings of a workshop held 7-12 July 1988, Saly, Senegal. Technical Centre for Agriculture and Rural Cooperation, Ede-Wageningen, Netherlands.

Starkey, P. H. and Goe, M. R. 1984. Report of the preparatory FAO/ILCA mission for the establishment of a TCDC network for research, training and development of draught animal power in Africa. AGA Consultancy Report, FAO, Rome.

Starkey, P. H. and Goe, M. R. 1985. Report of the third joint FAO/ILCA mission to prepare for the establishment of a TCDC network for research, training and development of draught animal power in Africa. AGA Consultancy Report, FAO, Rome.

Starkey, P. H. and Ndiamé, F. (eds) 1988. Animal power in farming systems. Proceedings of a networkshop held 17-26 September 1986 in Freetown, Sierra Leone. Vieweg (for German Appropriate Technology Exchange), GTZ, Eschborn, Germany.

Starkey, P. H., Sirak Teklu and Goe, M. R. 1991. Animal traction: An annotated bibliographic database. International Livestock Centre for Africa, Addis Ababa, Ethiopia.

Address

Paul H. Starkey, Oxgate, 64 Northcourt Avenue, Reading RG2 7HQ, United Kingdom.

Networking for Low-external-input and Sustainable Agriculture: The Case of UPWARD

Gordon Prain, Virginia N. Sandoval and Robert E. Rhoades

Introduction

Starting in the mid-1960s, the international agricultural research centres produced technologies that were to revolutionize agriculture worldwide. The new semi-dwarf crop varieties they developed were designed to use higher levels of external inputs to produce significantly increased yields. In addition, the new varieties were less sensitive to photoperiod or day length, allowing faster maturation and hence double or triple cropping. Although the introduction of modern high-yielding varieties (HYVs) has had a major positive impact on the food economies of many developing countries, it has also had some serious negative impacts on local natural resources and on the equitable distribution of wealth and benefits. More intensive cropping has rapidly exhausted the soil in many areas, leading to spiralling fertilizer use. Varietal uniformity among the HYVs, though increasing resource use efficiency in technological packages, has also made production more susceptible to climatic extremes and to attack by pests and pathogens, leading to the ever-increasing application of pesticides.

Richer farmers in the better endowed areas have been best placed to shoulder the increased input costs and thus benefit from the increased yields, even though their profit margins too have been gradually declining as increased inputs have been required to produce the same yields.

Poor peasant farmers, who make up the majority of the developing world's agricultural population, have fared less well. In marginal upland areas the common species grown, such as root crops and coarse grains, have yet to have their Green Revolution, although higher-yielding varieties are now available for some crops. Those growing traditional varieties of rice and wheat have often been placed at a double disadvantage compared with their richer neighbours: lower production which sells for a lower price per unit.

The failure of so many to benefit from the first Green Revolution was a clarion call for a second, a revolution not necessarily of technology but certainly of attitudes—one which turned upside down the conventional perspectives, philosophies and methods of the agricultural development community. The second revolution, now under way, turns from technology-driven development to farmer- or household-responsive development, from the supremacy of Western scientific thought to an adaptive logic which is sensitive to skills, opportunities and constraints on the farm, in the market-place and amongst consumer as well as producer households, from a short-sighted focus on productivity at any cost to a more holistic long-term appreciation of the 'user' (the person or group that will utilize the technology or management practice), the 'non-user' (the person or group who may not have access to the technology but will nevertheless be affected by its use) and the 'used' (the natural resources on which production depends).

UPWARD, the Users' Perspective with Agricultural Research and Development Network, is part of this second revolution, which has been building up steam over the past decade under such banners as Farming Systems Research, Farmer-back-to-Farmer, Farmer First, Diagnosis and Design and Farmer Participatory Research. These different initiatives have increasingly insisted on the logic and ethics of involving, at the base and in the beginning of research and development initiatives, those rural people most familiar with local needs and opportunities and with most to gain—and lose—from them.

Despite the many important reversals of thinking ushered in by the client orientation of the second revolution, there remained a strong production focus and a determined attachment to 'the farmer'—frequently the middle-aged male farmer at that—as the user-participant of natural choice. A key element of the original proposal which launched UPWARD was that in the vast majority of agricultural enterprises in the developing world it is not the individual male entrepreneur which is the principal social unit involved, but the household, with its web of gender and other role divisions and relations. Households, and complex agricultural systems in general, involve interactions and interdependencies along the food chain, linking production, conservation, distribution and consumption activities. A narrow focus on production is potentially misleading under these circumstances. The presence of these different roles and relationships in the food system requires us to broaden our concept of user and our approach to agricultural research.

In the new approach, indigenous knowledge and practices in different spheres of production, processing and consumption are taken as starting points, to be understood and built upon in a partnership between user and researcher. Through this partnership, relevant research priorities and plans can be designed and interdisciplinary research and development undertaken.

The Development of UPWARD

A paradoxical birth

UPWARD was born with a sweet potato in its mouth. Though more focused on a philosophy than a commodity, UPWARD was fathered by the International Potato Center (CIP), whose mandate is to solve priority problems that limit potato and sweet potato production and consumption in developing countries.

Many of the international agricultural research centres are often referred to as 'germplasm centres' to highlight their production orientation and commodity focus. Together with a stress on commodity production, some of the centres have also been closely associated with the top-down, transfer-of-technology model that is said to be dominant in international and national agricultural research thinking. Yet as early as the late 1970s, CIP along with several other centres had recognized that there was more to the solution of the world's food problems than high-yielding varieties. Post-harvest, consumption and nutrition issues were beginning to be taken much more seriously. Centres devoted to livestock research, policy and management issues had been formed, and a farming systems approach to research, including the social science disciplines, had been widely adopted. These concerns led in turn to much closer working contact with women and with households as a whole, and to the realization that most of the target group of production-oriented research so far had consisted of men. These were the internal shafts of light which led to the formation of the UPWARD proposal in the mid-1980s. Of course, these recognitions reflected or refracted issues already well aired in the wider world. The women's movement was already well established by the mid-1970s, while the concept of a food system became a radical analytical tool in Latin America during the same decade.

The paradox of UPWARD's link at birth with a commodity dissolves altogether when one looks at the crop itself. Sweet potato is one of the developing world's most widely distributed and versatile crops, adapting relatively easily to a range of cropping systems, including those in marginal areas, often associated with very low levels of external inputs and frequently identified as a good candidate for diversifying and sustainably intensifying crop production systems. It is commonly found as an alternative staple in upland systems and home gardens as well as a diversification crop in lowland systems. The bulk and perishability of root crops and their vegetative reproduction raise a whole series of questions and problems related to household storage, processing, marketing and the maintenance of planting material. Sweet potato is also a nutritious crop, with a significant production of edible protein and important quantities of micro-nutrients, and is thus potentially important for improved household nutrition. Yet of all the major food crops, sweet potato has probably received least research attention, perhaps because of its low status as a food.

Making a good idea work

With funding secured from the Government of the Netherlands, UPWARD was formally launched in April 1990 in an inaugural meeting attended by actual and prospective network members. The participants represented what could be called a cross-section of the world of agricultural research and development. The bulk consisted of researchers based at national agricultural research institutions, doing applied or adaptive research (or both) in the natural or social sciences. But there were also professors, research administrators, representatives of non-government organizations (NGOs) and funding agencies, and policy-makers. They came from all over Asia, bringing along their distinctive cultural, as well as intellectual, orientations.

The inaugural conference activated the network. While Plucknett and Smith's well-known article of 1984 on networking in international agricultural research emphasized the need to safeguard the calibre of participants, UPWARD chose instead to bank more on the promise and interest of researchers animated by the ideas aired during the conference or drawn into the UPWARD orbit through links with those who had been there. Through a willingness to support young researchers without any established track record, UPWARD sought to encourage innovative research and new approaches to agricultural problems, unhampered by mainstream premises and assumptions. As a consequence, the network had, and still has, a freshness about it that is difficult to find elsewhere.

During the inaugural conference, participants helped shape the general orientation of UPWARD through working groups on organizational structure, research projects, methods development, and training and information dissemination (UPWARD, 1990). Their deliberations and recommendations helped guide the coordinating office, both in reviewing research project proposals and developing a training programme. The guidelines relating to research retained considerable flexibility, so as to maximize innovation and experimentation.

The inaugural workshop also helped consolidate the coordinating office. UPWARD activities were initiated by CIP through its senior agricultural anthropologist Robert Rhoades, who became the founding coordinator. He established the office in Los Baños, the Philippines and hired the assistant coordinator and a small clerical staff shortly before the inaugural conference, which they helped organize. A Dutch nutritionist was also assigned to the office by the Dutch Government at about this time. A training assistant from the local university joined towards the end of 1993 to help organize the workshops, courses and training materials recommended by the conference. Finally, following a recommendation of the conference working groups to seek input from senior Asian scientists into UPWARD's orientation and research activities, Dr. Gelia Castillo, Professor of Rural Sociology at the University of the Philippines, who had already been involved informally in the early stages of UPWARD, was formally brought in as a senior consultant.

Following another proposal of the first conference's working groups, a second conference was convened in April 1991 to review research progress and bring together young UPWARD researchers and seasoned social and biophysical scientists (UPWARD, 1991). Working-group sessions held during the conference clarified the overall mission statement and the goals and priorities of the network's four principal research areas: production systems, genetic resources, marketing, processing and consumption, and policy and institutional issues. Following subsequent review, policy and institutional issues have now been integrated as key research aspects of the other three areas, rather than being a separate research area.

Networking a philosophy

One of UPWARD's earliest tasks was to stimulate the diffusion of user-focused, food systems-oriented research approaches within agricultural research and development institutions in Asia. To begin this diffusion process, the nascent links developed through the inaugural conference were activated through follow-up. Ideally, a development-oriented network is a set of mutually supporting, evenly weighted links between people and institutions with common interests rather than a nucleated entity radiating outwards, but a certain heaviness at the centre is inevitable during the process of network formation.

UPWARD therefore supported the development and spread of user-sensitive methods not only by providing expertise from its coordinating office but also by setting up connections between experienced practitioners in one set of institutions with actual or potential UPWARD researchers without previous experience based in other institutions. The first event sponsored was a Philippines in-country training course for prospective UPWARD project leaders. The trainers were social scientists from metropolitan and provincial universities and the UPWARD assistant coordinator. The trainees came mostly from government research institutes and the technical departments of provincial universities. A second, international training course on farm household diagnostic skills was led by the UPWARD coordinator and staff, drawing on support from the local university. These initial courses aimed to 'train trainers'. Several participants subsequently served as trainers in 'echo seminars' which they organized and UPWARD funded in different parts of the Philippines, broadening existing network links and stimulating the formulation of research proposals, as well as bringing new inquiries about joining the network. The modifications which these echo-trainers introduced into the content and format of the training events, to suit their problems and needs as well as to reflect the feedback given to coordinating staff, has helped to improve subsequent workshops and courses.

Through feedback and direct observations, it is apparent that the echo-seminar idea has tremendous potential for spreading approaches and methods very far and fast. Nevertheless there is a need for ongoing support to 're-echo' the messages among the same participants over time. We need to bear in mind that courses and seminars create a rather special context, in which common goals, joint activities and the resulting group spirit quickly generate great enthusiasm. Once participants return to their habitual environment, however, more conventional approaches to research—especially when practised by superiors—can easily undermine that enthusiasm. Repeated modest courses and continuing contact are very much needed.

Network research

Since the aim of the methodology development discussed in the previous section is to help break the mould of conventional research by equipping researchers with greater sensitivity to users, an eclectic strategy was adopted in the identification of researchers and in the evaluation and acceptance of proposals during the early stages of network development.

The researchers are predominantly young, almost half of them women (Table 1), and they are based in universities and extension services as well as in national agricultural research institutions. There is a very wide disciplinary spread, fairly evenly distributed between the socio-economic and the biophysical sciences (Table 2).

Table 1 Distribution of UPWARD researchers by country

Country	Male	Female	Total
China	2		2
Indonesia	8	3	11
Nepal	3	3	6
Philippines	10	18	28*
Sri Lanka	3	1	4
Vietnam	2	1	3
Total	28	26	54

* Includes UPWARD's Dutch Associate Expert and three visiting graduate researchers/thesis candidates from the Netherlands.

Research proposals were solicited through personal contacts and through the distribution of flyers which described possible research areas, potential funding sources and the structure of proposals. As already mentioned, the inaugural conference was the catalyst for the development of a Philippine research cadre and the funding of two multiple activity projects. The methods training events were also fertile grounds for project development: of the 97 people who have attended, 56 have implemented or will implement UPWARD projects. Proposals have been evaluated by the coordinating office, the senior consultant and the project leader of UPWARD research in the Philippines. Depending on the type of proposal, additional opinions are sought from specialists at the local university, CIP staff or, occasionally, researchers associated with Wageningen Agricultural University, where an informal Dutch support group gives information and other kinds of back-up to UPWARD. For UPWARD's second phase a more systematic form of proposal evaluation is envisaged, comparing different projects using an interdisciplinary team of network members and coordinating office staff. Efforts will be made to gauge the project's potential to contribute to the network and to the originating institution. Attempts will be made to improve consultation with the proponents of proposals which show promise and demonstrate empathy with a users' perspective.

Table 2 Disciplines involved in UPWARD's research

Project	Social science disciplines	Biological science disciplines
China:		
Potato production, consumption and marketing in Zhejiang Province	Extension	Systems agronomy
Indonesia:		
Farmers' perspectives on potato technologies	Human ecology	Entomology, plant pathology, agronomy
Sweet potato processing in West Java	Economics	Physiology, breeding, post-harvest processing
Collection of land races and associated indigenous knowledge in Irian Jaya	Socio-economics	Agronomy
Nepal:		
Sweet potato in Nepalese food systems	Extension	Forestry
Importance of sweet potato in household economy and consumption	Extension	Soil science
Philippines:		
Local adaptation of technology in sweet potato production, distribution and consumption (5 sub-projects)	Family resource management, extension, sociology, nutrition	Environmental science
Philippine sweet potato sub-system projects (8 sub-projects)	Economics, extension, rural sociology, development communication, accounting	Environmental science, agronomy, horticulture, entomology, weed science, plant breeding, food technology
The role of sweet potato in the diet of pre-school children	Nutrition	
Sweet potato consumption in two villages in Cordillera	Nutrition	
Farmers' indigenous knowledge of sweet potato production in the Cordillera	Rural sociology	
Memory banking of indigenous technology associated with traditional crop varieties	Anthropology, human ecology	

Table 2 continued

Project	Social science disciplines	Biological science disciplines
Philippines (cont.):		
Indigenous knowledge of pests among sweet potato users	Socio-economics	Entomology, virology, weed science, plant pathology
The Aetas world view of sweet potato culture: guide to their resettlement and rehabilitation	Cultural anthropology	Agronomy, plant breeding
Market assessment study for sweet potato processed products	Human ecology, family resource management	Agronomy
Needs and capability assessment for sweet potato industry development among farm households	Human ecology	Agronomy
Knowledge systems analysis of sustainable agriculture in the Philippine uplands	Development communication	Breeding, agronomy, forestry
Sweet potato home gardening technology development	Socio-economics, nutrition	Agronomy
Collection of germplasm and associated indigenous knowledge in Ifugao: a preliminary study		Environmental science
Ethnographic video documentation of upland farmers' adoption of sweet potato technologies	Development communication	
Sri Lanka:		
Socio-economic and ethnobotanical baseline study of sweet potatoes	Economics, extension	Plant breeding
Thailand:		
Thai food habits and potential for sweet potato processing	Home economics/extension	
Sweet potato production in Thailand	Home economics/extension	
Vietnam:		
Sweet potato utilization in Vietnam		Food technology
Development and diffusion of processing technology for sweet potato transparent noodle production	Food technology	

Network researchers and network associates

Like many networks UPWARD quickly built up a hard core of active members close to the hub, mostly consisting of the trainees turned trainers and of those receiving support for research projects. This core has been predominantly Filipino but evenly spread across genders and disciplines. NGO participation has so far been mainly indirect, through university faculty.

When networks become closed systems they turn into cliques or in-groups, often distancing themselves from broader national or regional decision-making processes. UPWARD recognized the need for a dual orientation to develop human capacity internally *and* influence research directions and public policy externally. Equally important, it saw the need to maintain an open flow between these two orientations. A subnetwork was therefore established consisting of a pool of friends, sympathizers and opinion and policy-formers, principally in Asia but also in other parts of the world (Table 3). To energize the links with these associates as well as with the network's researchers, information is disseminated in several forms (Table 4). *Notes from the Field,* UPWARD's newsletter, describes network activities and research results and announces plans and recent library acquisitions. A recently launched working paper series disseminates and seeks comments on reports of selected research activities. A proceedings series publishes the collected papers and discussions from meetings and workshops. In a network as heterogeneous as UPWARD it is often difficult to achieve the brevity and standardization required in proceedings. Nevertheless, through a combination of mutual respect and the talents of participants, the two proceedings of the UPWARD annual conferences (1990 and 1991) already published managed a degree of coherence, which it is hoped the three workshop proceedings currently in preparation will build on.

In a still broader effort to influence agricultural research and development, UPWARD has helped sponsor international conferences on Asian farming systems and home gardens, and has supported the award of prizes at such conferences for best papers with regard to scholarship and the embodiment of a participative, user-sensitive philosophy.

Achievements and Limitations

In its progress from the seeds of an innovative idea within a forward-thinking agricultural research institution to its emergence in Asia as a fledgling network of researchers dedicated to turning agricultural research upside down, UPWARD has some real achievements to its credit. It has helped to create a continuing dialogue between international and national scientists, between natural and social scientists, and between seasoned and young researchers. Workshops have exposed national technicians, natural scientists, and more quantitatively oriented social scientists to alternative, 'softer' approaches to research on agricultural problems. At the same time, social scientists have acquired considerable technical knowledge through the participation of biophysical

Table 3 Distribution of UPWARD network associates*

Region/country		No. of associates
Asia		141
Bangladesh	5	
Pakistan	3	
Bhutan	2	
Philippines	43	
China	3	
Sri Lanka	11	
India	20	
Taiwan	1	
Indonesia	15	
Thailand	22	
Nepal	10	
Vietnam	6	
Africa		10
Latin America		10
Europe and North America		56
Total		217

* Network associates are those people and institutions interested in the issues and activities with which UPWARD is concerned, who receive the newsletters and other publications and who supply news and literature from their own programmes.

scientists in workshops and particularly through their increasingly close contact with sophisticated local users of technology. With increasing intensity of these types of exchanges, research becomes more truly inter- rather than multidisciplinary. An entomologist is currently leading an interdisciplinary team looking at the indigenous classification of sweet potato pests and how to manage them; an agronomist, with support from the Health and Social Welfare Departments of the Philippines, is promoting pot culture of sweet potato cuttings as a source of green vegetables for urban slum dwellers; an anthropologist, with support from plant breeders, is designing 'memory banks' based on farmers' indigenous knowledge of traditional varieties, to complement gene banks.

Thankfully, UPWARD has been allowed considerable latitude in the development of new links, especially in the search for new research areas, and this accounts for the diversity of innovative projects now under way. Within the bounds set by donor and auditor requirements, the organization of the network has remained quite simple, in terms of the requirements placed on project proposals, funding arrangements and the size and functioning of the coordinating office.

Table 4 Distribution of UPWARD publications, 1990-92

Publication	Number of copies	
	Asia	Rest of the world
Methods manuals:		
Training of trainers	325	143
In-country manuals	323	129
First annual proceedings	202	98
Second annual proceedings	78	172
Best paper award	137	68
Newsletters (3 issues)	656	539
Calendar	244	156
Working papers 1-4	(in press)	

However, if we are to characterize the network in terms of anthropological network theory (Bott, 1971) or along the lines more practically laid out in the excellent paper by Fernando (1989), UPWARD is still a 'loose-knit' or 'low-density' network. That is to say, most of the active links are with or through the centre (the coordinating office). As mentioned earlier, all networks can expect to have a 'heavy' centre to begin with, but understanding the nature of this 'heaviness' can help to transform the network into a more participative entity (Figure 1).

Although UPWARD network members share many common interests, these are in most cases latent at present, remaining to be fully expressed and developed. There is an especially low level of communication between network researchers and network associates. Furthermore, communications are still too frequently one-way, from the centre. One reason for this is the funding role performed by UPWARD, which accounts for a very large share of the communications traffic. But the main reason is the continuing attempts to build and consolidate from the coordinating office. This brings to mind the image of a centre and its 'satellites', a closed system which, unless transformed, could lead to entropy and marginalization. Networks must be open, and techniques need to be identified which support such openness.

Geographically, the network is heavily weighted towards Filipino researchers and associates, a tendency mainly attributable to the influence of the coordinating office. Though it is entirely reasonable for a newly established network to begin by forging links close to home, there is a clear need to redress the geographical balance in subsequent phases.

Figure 1 Current links of UPWARD

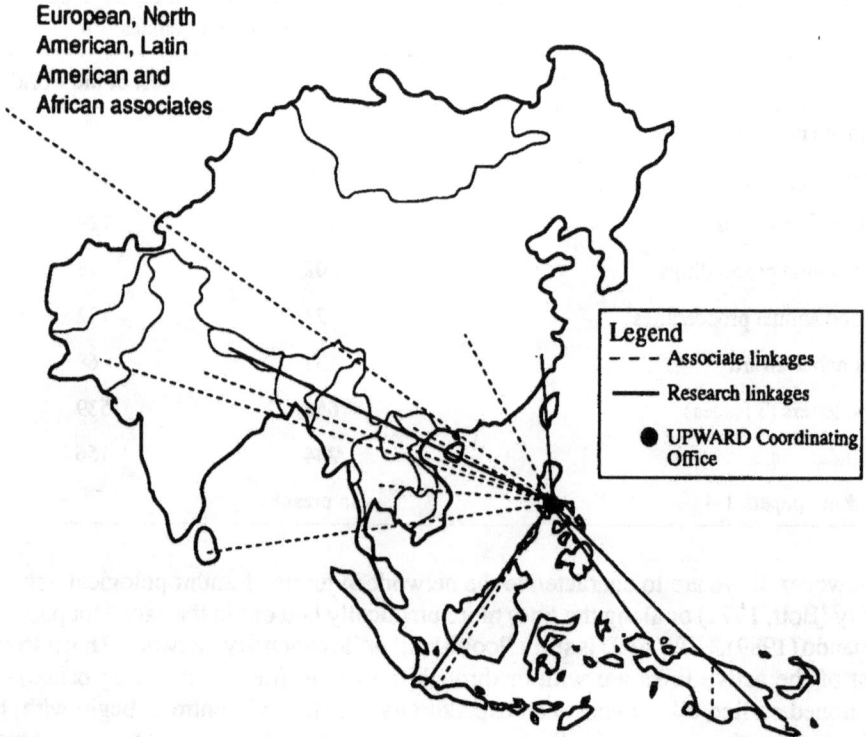

Perhaps a more serious barrier to the long-term institutionalization of the ideas and approach of UPWARD is the fact that most network members are still individuals rather than institutions. Again, the reasons for this are clear. UPWARD's early priority was to identify new blood in the research and extension systems of the region and to break free from the conventional institutional and bureaucratic structures which dominate commodity research. Research support to university teachers and other researchers has often led to *de facto* agreements with their institutions. However, these need to be more formally established to facilitate greater acceptance and diffusion of the approach.

Looking Ahead: Towards a Denser Network

Some of the needs expressed in the preceding section are already being addressed. Many informal contacts with institutions have been firmed up and the institution and its library are now the recipients of UPWARD publications as well as, or instead, of the individual researcher. The range of institutions with which UPWARD is linked is also being broadened, with special emphasis on NGOs, both through informal contacts and cooperation and through the funding of NGO projects.

Several initiatives are planned for the second phase, to speed up the transformation of UPWARD towards a denser, more tightly knit network (Figure 2). To redress the geographical imbalance, a more equitable distribution of projects is planned. At the same time the network will aim to develop autonomous links, between researchers in particular localities, regions or countries. This will involve the initial UPWARD network member acting as a kind of 'animator' (Fernando, 1989) in his or her area, especially linking colleagues and senior staff to the rest of the network. Where two or more network members reside in the same region or country, this helps both the network and the institutionalization of a user's perspective. The support of new research projects will be partly guided by the important criterion of establishing a critical mass of network members by building these autonomous links.

Figure 2 Future links of UPWARD

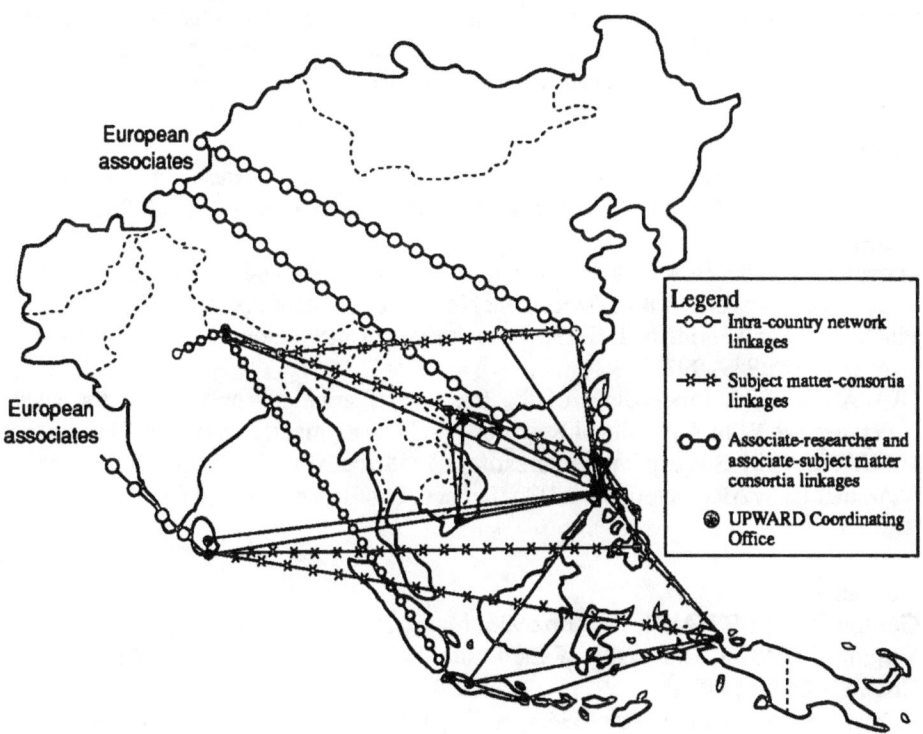

To further stimulate horizontal links across the Asian region, UPWARD will facilitate the establishment and functioning of sub-groups or subject matter consortia on specific topics, such as gender issues or livestock-crop interactions. One such consortium, on user perspectives in genetic resources research, has already been established through an international workshop. It links Asian researchers with funding and research institutes in the North.

One set of network associates of particular importance is known as 'the Dutch Support Group'. This is an interdisciplinary group of faculty from Wageningen University whose role is to provide inputs on research activities and directions, to offer support in the area of methodology and, in some cases, to participate in training events. Using modest funds now available from the Dutch Government, an important role of this group's members in the future will be to support the subject matter consortia appropriate to their area of specialization.

Networks are open systems and are therefore subject to myriad influences in the way they develop. This is as it should be. UPWARD's aim is to catalyze the expansion of autonomous research and development capacity and activities within the framework of a participative, user-focused philosophy. We hope to achieve this by providing modest funding, stimulating the flow of information, supporting workshops on methods and approaches and, finally, disseminating the goodwill and the shared belief that, through connecting in these ways, we can make a difference.

References

Bott, E. 1971. *Family and social network: Roles, norms and external relationships in ordinary urban families*. 2nd ed. Social Science Paperbacks. Tavistock Publications, London.

Fernando, S. 1989. How networks function: Some structural and international aspects of the IRED Network in Asia. Document No. 3, Occasional Papers. IRED, Colombo.

Plucknett, D. and Smith, N. J. H. 1984. Networking in international agricultural research. *Science* 225:989–993.

UPWARD. 1990. Proceedings of the Inaugural Planning Workshop on the User's Perspective With Agricultural Research and Development. Los Baños, Philippines.

UPWARD. 1991. Sweet potato cultures of Asia and South Pacific: Proceedings of the 2nd Annual UPWARD International Conference. Los Baños, Philippines.

Addresses

Gordon Prain, UPWARD, P.O. Box 933, Manila, Philippines.
Virginia Sandoval, University of California, Board of Environmental Studies, Santa Cruz, 95064 CA, USA.
Robert Rhoades, University of Georgia, Department of Anthropology, Athens, 30602 GA, USA.

PART FIVE

Support Organizations

Research Networks: Evolution and Evaluation from a Donor's Perspective

Terry Smutylo and Saidou Koala

Introduction

There is general agreement today on the importance of agricultural research in economic and social development. Nevertheless, agricultural research continues to receive low priority in the allocation of financial resources by governments. In this climate, agricultural research organizations must not only be efficient but be seen to be so. They must demonstrate the value of their work. They must be able to prove in facts and figures that they are adequately repaying the societies that invest in them. According to Gastal (1987), any possible means of increasing observable benefits at lower cost should be pursued. Research networks are such a means. Such networks are playing an increasingly key role in the coordination of international efforts to develop improved technologies for food production, especially for small-scale farmers.

An agricultural research network can be defined as a voluntary association of research organizations with sufficient common objectives to be willing to adjust their research programmes to, and invest resources in, associated activities in the belief that they will thereby meet their objectives more efficiently than if conducting their research alone (Banta, 1982). In setting up a network, three basic functions are typically required. A planning function brings agreement on the objectives of the network and the relative priority to be attached to them; a cooperation function allocates resources to the activities required to meet the objectives; and a coordination function organizes the activities of the participants to achieve the objectives efficiently.

This paper reviews the involvement of the International Development Research Centre (IDRC) of Canada in network support. Using data from informal and formal evaluations, it also assesses the performance of these networks and presents the lessons learned.

Evolution of IDRC Support

The majority of IDRC-funded networks have evolved from individual projects in different countries. Often a prerequisite to network formation, these early project experiences helped determine the degree of common interests and problems among participating countries. Awareness of shared problems enables the new network to reach agreement on objectives and priorities, allocate resources, assign responsiblities and organize activities cost-effectively.

Thus, IDRC has emphasized networks as a means of linking scientists working on similar problems in different countries rather than as mechanisms to foster or fund research in countries. It is an approach which has given IDRC a great deal of flexibility in responding to the expressed needs of developing country scientists as well as to fiscal pressures at home to deliver better technical and material support while reducing overheads.

Network members share information, technologies and methods, pooling their efforts to solve problems of mutual concern. Over the years, both participants and evaluators have found the networks supported by IDRC to be effective in generating and sharing knowledge about development, and IDRC has come to see networking as indispensable in the pursuit of efficient scientific research and technology adaptation. Involvement with networks has grown from an average of 13 network projects per year in the 1970s (9% of annual appropriations) to 79 networks and/or network projects per year in the 1980s (24% of annual appropriations).

Types of Network Supported

In general, the way networks are classified depends on the purpose of the classification. Those interested in dynamics have classified networks according to the level of integration of the different actors (Banta, 1982); others have classified them on the basis of commodity, production system or discipline (Faris, 1991). A widely reported classification (World Bank, 1987; Faris and Ker, 1988; Faris, 1991; Valverde, 1988) is the one proposed to the Special Program for African Agricultural Research (SPAAR) by Ralph Cummings Jr. and Calvin Martin (SPAAR, 1987). Based on the level of research in the network and the degree of collaboration used to plan and conduct research, this typology is further described in the paper by Plucknett et al in this book (p.187).

IDRC is currently using a classification based on what is exchanged relative to particular development problems or needs. According to this classification it is supporting four basic types of network:
° Horizontal networks linking institutions with similar interests working in the same or a related field.
° Vertical networks of institutions working on different aspects of the same problem or on different but interrelated problems.
• Information networks providing centralized information services to members and other users, enabling them to exchange information as needed; and
• Training networks, which provide training and supervisory services to participants working independently in their own research areas.

Across these four general categories there is wide variation. Networks evolve according to members' needs, the resources available and the kinds of contact established. Most of the early agricultural networking fostered by IDRC was a response to the isolation faced by developing-country scientists in the 1970s and to the need for critical masses of scientific effort if progress was to be made. Many of these networks started as informal groupings following workshops or conferences, moving to more formal associations with resources for coordination or a secretariat, as common interests and agreements to share resources were established. A few formed a nucleus around which a new commodity research institute crystallized.

Patterns of Support

Between 1970 and 1991 IDRC invested close to C$ 242 million in network-related projects. Some of this went directly to the creation and coordination of networks; some funded research by network members. Figure 1 shows these expenditures by year for the period. Looking at distribution by sector, agriculture clearly led the way in the use of research networks. Sixty-two percent of network-related funding was for agricultural sciences, 12% for social sciences, 12% for information sciences, 11% for health sciences and 3% for earth and engineering sciences (Figure 2). The geographical distribution of IDRC-supported network projects is shown in Figure 3. Latin America has been by far the most active region, with 39% of the projects, followed by Africa, with 28%. Network activity in Asia and the Pacific has been the lowest over the 20-year period, but has increased significantly over the past 2 years. Global networks are those which include more than one of these regions.

IDRC's substantial experience with networks, some of which has been formally documented in evaluations and staff papers, has led it to recognize networks as an important way of organizing resources for development-related research. In particular, the viability and usefulness of networks have increasingly benefited from enhanced access to new information technology. Yet there remain a number of concerns about networks which need to be addressed as we adapt this mechanism to present and future needs. We are approaching 'network overload' in some subjects and geographical areas. Donor coordination, itself a form of networking, is often weak. It is not easy to recognize and foster the appropriate conditions for network formulation, sustainability of dissolution, or to ensure that network resources are used efficiently and effectively. The tools needed to measure network performance are not well developed. There are relatively few multidisciplinary networks which operate effectively. Network links with national research and development systems are often weak. One way of bringing about improvements in efficiency is to study different experiences across countries, regions and subject areas. Through these studies principles for the design and management of networks can be deducted, and then applied to individual networks.

Figure 1 Number and funding level of network projects supported by IDRC, 1970-91

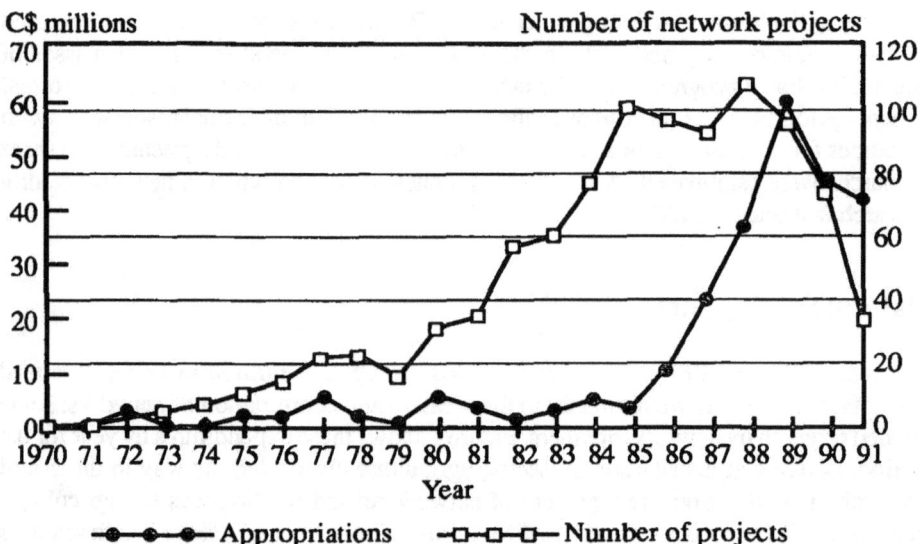

Examples of IDRC-supported Networks

The following examples show how some IDRC-supported networks were formed and have evolved.

Oilseeds network

Oils and fats are essential components of the human diet. Nutritionists recommend that about 20% of energy requirements come from oils and fats, which are concentrated forms of energy allowing efficient utilization of fat-soluble vitamins. Requirements for energy vary with age, weight and the level of physical exertion, but an average adult requires a minimum of about 55 g of oil or fat per day, or 20 kg per annum.

In many developing countries average consumption is much lower than this, often varying from 1 to 10 kg per annum. Low oilseed production is thus a major cause of the protein–energy malnutrition which affects enormous numbers of people in these countries.

Since the late 1970s, IDRC has supported oil crop improvement projects in China, India, Sri Lanka, Nepal, Pakistan, Mozambique, Tanzania, Ethiopia, Sudan and Egypt. These projects have included work on groundnut, brassica, sesame, sunflower, safflower, linseed, niger seed and castor. Among these commodities, only groundnut has been the responsibility of an international agricultural research centre, namely the International Crops Research Institute for the Semi-arid Tropics (ICRISAT).

As a group the oilseed crops are important, but individually most are neglected minor crops. Pulling them together through a network provided a substantial intellectual impetus to research and the necessary basis for the more cost-effective use of research resources.

IDRC support has been critical in focusing research attention on these crops, particularly in South Asia and Eastern and Southern Africa, where the crops are grown both for home consumption and as a source of cash income, often by very poor people. The networking approach has been particularly important in strengthening links among oil crop researchers in Canada, Asia and Eastern and Southern Africa, and between stronger and weaker programmes working on the same crop. The absence of an international agricultural research centre with responsibilities for most of the crops involved has made network activities more difficult to plan and backstop, but has also heightened their importance.

West African Farming Systems Research Network

This network emerged from the perception that farming systems research (FSR) had considerable potential for improving small-scale farming in West Africa, where severe problems were seen to be associated with the breakdown of existing farming systems caused by increasing population pressure on the land (Koala and Banta, 1989).

The West African Farming Systems Research Network (WAFSRN) evolved from a professional society formed by scientists. Its primary objective was to assist national practitioners seeking to improve their farming systems approach to agricultural research

Figure 2 Percentage network project appropriations to different scientific sectors, 1971-91

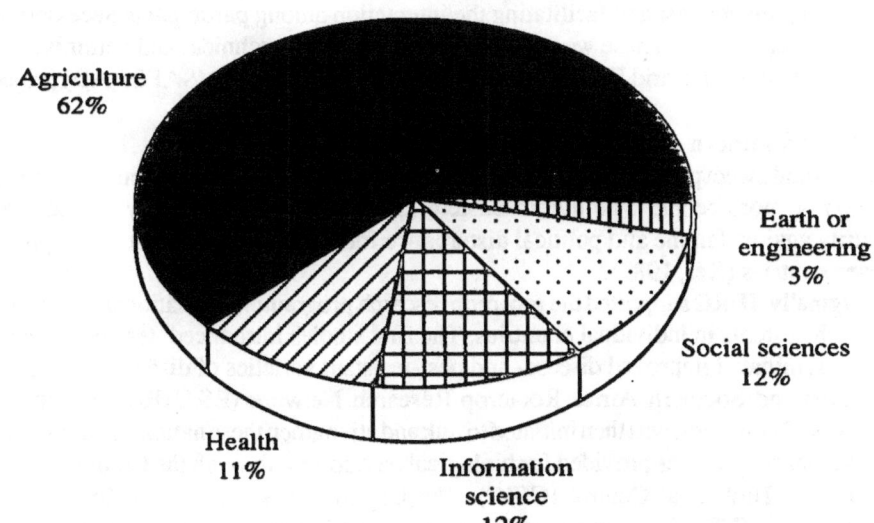

Figure 3 Percentage network project appropriations by region, 1970-91

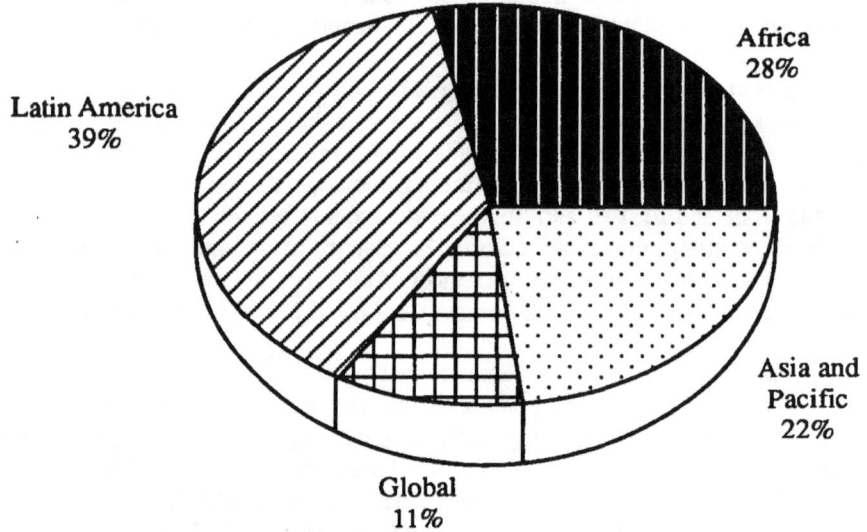

and development. The network fostered the exchange of relevant experience among researchers, collected information and made it available to members, and promoted training programmes in FSR methodology.

The provision of a full-time coordinator and support services proved critical in stimulating the interest and facilitating the interaction among participants necessary for effective networking. These were strengthened further by technical and administrative inputs from the Semi-arid Food Grain Research and Development (SAFGRAD) project.

East and Southern Africa Rootcrop Research Network
Cassava and sweet potatoes are important staple foods and cash crops for many of the 200 million or more people who live in Eastern and Southern Africa. They provide food security against famine and political disturbance for many of the poorest of the poor in these countries (Ker, 1989).

Originally, IDRC supported the root crop research programmes of national agricultural research systems in individual countries. The first studies introduced, tested, evaluated and disseminated improved disease- and pest-resistant varieties of different root crops. The East and Southern Africa Rootcrop Research Network (ESARRN), which now numbers 12 countries, was then initiated to link and strengthen these national programmes. Additional support was provided for biological pest control through the Commonwealth Institute of Biological Control (CIBC). The United States Agency for International Development (USAID) began supporting the network a few months after IDRC. The network was administered through the International Institute of Tropical Agriculture (IITA), which provided additional support and an IITA scientist as network coordinator.

ESSARN is an example of a horizontal network supported by more than one donor agency that is linking institutions with similar interests working on the same crops.

Latin American Aquaculture Network

Aquaculture development in Latin America has been promoted since the early 1980s through a number of IDRC projects in Brazil, Colombia, Panama, Peru, Chile and Ecuador (Buzeta, 1989).

Aquaculture has been recognized by all Latin American countries as an important production alternative, contributing to local protein supplies as well as to cash income for poor people.

The need to establish coordinating mechanisms to promote regional cooperation and strengthen national scientific and technological capacities has been stressed at several international meetings held to analyze aquaculture development in the past 15 years (Buzeta, 1989). In consequence a network has now been initiated with the general objective of contributing to the research, planning and training capacities of the participating aquaculture centres.

An important feature of this network is its focus on strengthening the technology transfer capability of aquaculture centres, through training and information exchange. To facilitate the exchange of information between researchers and producers, integrated planning/action groups were formed at the community level (Davy, personal communication) involving producers, researchers and representatives of government and non-government organizations, who together carried out the project's research and development activities.

The Latin American Aquaculture Network (LAAN) provides an example of a network seeking to involve end users in the research and development process. By involving the producers from the outset the project hoped to make them feel a greater sense of ownership of the technology developed, thereby facilitating technology transfer.

Need for Monitoring and Evaluation

Networks have proliferated over the years, as participants and donors have come to see them as a means of achieving more cost-effective use of resources, a more innovative approach to research, greater chances of impact and enhanced capacity building and institutional development. Various publications have discussed these benefits and the principles for successful networking (Faris, 1991; Plucknett et al, 1990). However, networks do not always live up to the expectations of their members. Monitoring and evaluation are essential, both to improve individual networking and to compare experiences and disseminate management lessons more widely.

Unfortunately, current literature on networks reveals little information on the methods available for evaluating them. The most significant contributions are from Valverde (1988) and Faris (1991), both dealing with internal (monitoring) and external evaluations.

Internal evaluation (monitoring)

It should be the responsibility of every network to evaluate its own activities, to identify and build on strengths and deal with problems as they evolve, before they become serious. Among the many possible internal evaluation methods, Faris (1991) has suggested four that are effective, involving coordinators, steering committees, network-wide workshops and monitoring tours.

Centralized coordination is a key to good network management. The coordinator's job is to supply technical and moral support to the national programmes, to help establish effective operational procedures (technically, financially and administratively), to act as a communications link among the members and as a buffer between conflicting national programme interests, to organize network activities, and to provide leadership. In his or her daily dealings involving these activities, the coordinator also monitors and evaluates operations, building on successes and checking for potential problems.

A steering committee has a monitoring and evaluating role more formal than that of the coordinator, often dealing with the same issues, but as they affect policy or strategy for the network as a whole.

IDRC's experience suggests that workshops and the publication of workshop proceedings can be effective as evaluation mechanisms, provided they are organized with good representation of the membership and periodically review the network's goals, mandate and objectives as well as its research priorities.

Monitoring visits by the coordinator, selected members or donor's representatives may be considered as a form of internal evaluation and can be a good way of identifying problems. The assessment criteria to be used during such visits will depend on the goals or the interests of the visitors. Donors, for example, may want to know how the network has increased the cost-effectiveness of research, and may ask coordinators or evaluators to report on this. Thus the borderline between coordinating networks and monitoring them sometimes becomes blurred. The monitoring role of IDRC programme officers has been adapted to fit the particular coordination mechanism used. Staff with a keen personal interest in a network's area of research may become heavily involved in coordination as well as monitoring. More formal external donor evaluations usually include questions assessing the monitoring and coordination functions.

External evaluations

External evaluation is useful for providing network members and organizers with data on programme operations and impact compared with initial objectives. The expectation is that, by injecting new insights into network management and participation, performance will be enhanced. However, acceptance of evaluation findings is a key to their being used to make improvements. A major strength of internal evaluations is that network participants are involved in assessing their own activities. Allowing them to participate in external evaluations as well makes it more likely that the findings of the evaluation will be accepted.

The most comprehensive method for evaluating networks published to date is the one proposed by Valverde (1988) and reviewed by Faris (1991). It aims to identify and analyze the key constraints and elements that influence the execution of agricultural research network programmes. Through a systematic analysis of network elements, the method enables a list of weaknesses, strengths, threats and opportunities to be generated, leading to recommendations for appropriate adjustments. The Valverde method relies on both informal and formal data collection, varies according to the nature and type of network, and encompasses the assessment of biological research activities, regional exchange activities, and network management.

Evaluation Findings

Evaluations have been completed for 15 IDRC-supported networks since 1982. Each was undertaken to respond to particular information needs and addresses issues of significance to the network in question at the time. Although they were not designed to contribute to an overview of the lessons learned about networks, by comparing and synthesizing information from these evaluations it is possible to draw conclusions regarding their planning, organization and management. In addition, it is possible to examine how different types of network function—whether they have distinctive benefits, and their effects on the participants.

The 15 networks encompass three basic types of activity: research (9 networks), information exchange (3 networks) and technology transfer (3 networks). Research networks were designed to conduct basic or applied research on various topics, and involved universities, government and/or non-government organizations. Information networks focus on establishing systems to manage and/or analyze and exchange data, including the provision of bibliographic information which may or may not have been computerized. Technology transfer networks engage in the assessment and dissemination of improved technologies and technical skills. The networks ranged in age from less than 1 year to 12 years old at the time of evaluation. Some were specific to a region; others were global. Different disciplines were included in each. Table 1 provides the salient information on each network.

All of these evaluations were ex-post, in that they were started after the networks had begun operating. They were designed to provide information of importance to the specific network and therefore covered widely different issues. As a result it is difficult to aggregate or synthesize their findings. However, one can extract a number of useful lessons for improving network operations.

Table 1 Data on networks evaluated by IDRC

Name (year of evaluation)	Age (years)	Region	Discipline(s)
Research networks:			
Adolescent Fertility Network in West Africa (1992)	6	Africa	Social sciences
East and Southern Africa Rootcrop Research Network (ESARRN) (1992)	6	Africa	Agriculture
Latin America Aquaculture Network (LAAN) (1991)	5	Latin America	Aquaculture
Latin American Urban Hydrogeology Network (LAUHN) (1990)	0.5	Global	Hydrogeology
Macro-economic Analysis Program (MAP) (1985)	1	Africa	Economics
Network for Aquaculture Genetics in Asia (NAGA) (1989)	4	Asia	Aquaculture
Oilseeds Network for East Africa and Southern Asia (1992)	11	Latin America	Agriculture
Post-production and Food Industries Advisory Unit (PFIAU) (1988)	4	Africa	Agriculture
Red de Investigación en Sistemas de Producción Animal Latinoamerica (RISPAL) (1988)	7	Latin America	Animal science
Information systems networks:			
Agricultural Information Bank of Asia (AIBA) (1986)	12	Asia	Agriculture
Regional Information System on Planning (INFOPLAN) (1985)	10	Latin America	Economics and social science
IDRC Information Sciences Division, Caribbean Program (1989)	12	Latin America	Information science
Technology transfer networks:			
Caribbean Technology Consultation Services (CTCS) (1989)	3	Latin America	Industry
Rural Energy Technology Assessment and Innovation Network (RETAIN) (1989)	4	Global	Energy
Asian Network for Industrial Technology and Information Extension (TECHNONET Asia) (1982)	10	Asia	Industry

Viability

External resources are usually required to set up and coordinate network activities. In providing such support, donor agencies should recognize that a long-term funding commitment is required. Whatever the intentions of donor agencies with regard to the amount and duration of funding, the time period for which funding will be made available should be stated at the outset. This point was emphasized in half the evaluations. A fairly long time is needed to plan for the cessation of external support.

Many networks will be unable to maintain themselves financially without being offered at least some external support. While most networks have been able to attract external funding, only the technology transfer and information networks had generated what may prove to be more sustainable sources of financial assistance from participating organizations, clients and/or governments. As one might expect, these networks did not feel an urgent need to anticipate the cessation of donor funding, whereas the research networks were very sensitive to the issue of planning for the withdrawal of IDRC support in the future.

Adaptability

An important issue highlighted by the evaluations is the relationship between network coordination, control and ownership, and the degree of adaptability and responsiveness required by the network. Ineffective leadership and/or coordination, when combined with little or no local sense of ownership, tends to result in a network that does not adapt well to changing circumstances and is unresponsive to the needs of its members. The reverse is also true. Two evaluations serve to illustrate this point.

The first concerns the Agricultural Information Bank of Asia (AIBA). Initially there was poor leadership within the network, evidenced by 'insufficient assistance from the regional centre to the national centres'. There was also a feeling on the part of the national centres that they were not 'full and formal partners in policy making'. In other words they lacked a sense of control or ownership over the network at the local level. These two factors have led to the 'inability of the regional centre to adapt to the ever-changing situation of the national centres, which have developed fast'.

In contrast, in the case of the Latin American Aquaculture Network (LAAN), 'the role of the regional network coordinator in providing advice to research institutions and government policy-makers was seen as a major strength of the regional network'. At the national level, the Colombian work group took ownership of the agenda and began setting their own goals. The combination of strong network coordination and strong local ownership made the members feel that it provided appropriate information and a basis for action which suited their needs.

Information exchange

The advent of a network is often a response to the need to improve the exchange and dissemination of information (Akhtar, 1990). An important implication of this is the need to gear information dissemination to the needs of users, and to make them aware of the services available. From the evaluation data it appears that, ironically, the information and technology transfer networks were weak in disseminating internally derived information. They also appear to have had difficulty recognizing the information and service needs of their end users. This discrepancy requires further study, given that these networks tend to be more user-oriented than research networks, which direct their outputs mainly to other researchers and research institutions. The target groups of information dissemination in the different types of network are presented in Table 2.

Networks also serve to establish additional links between network members and other national, regional or international organizations. These include connections between researchers who have previously had little or no contact with each other, and between institutions and government or non-government agencies. Technonet Asia (TA) has been particularly successful in establishing links, as noted by the evaluation:

> Over the years, TA has developed active cooperation with many international organizations; apart from IDRC its parent, and CIDA its other principal donor, TA has now cooperation with more than 60 international bodies; this cooperation ranges from simple information and personnel exchanges to elaborate joint venture projects (Jarmai, 1982).

The technology transfer networks have been the most successful in establishing contacts between individual researchers, while the research and information networks seem to have concentrated more on institutional links. Only the research networks seem to have established links with government agencies and/or universities. Technology transfer and information networks tended to focus their linkage activities on non-government organizations. In terms of intra- or inter-regional links, research networks appear to be more likely than the others to seek to establish relationships with international, or at least extra-regional, institutions. This reflects their characteristic role in transferring strategic research techniques and approaches to developing countries.

Attention should be paid to the issue of who is to be the ultimate beneficiary of the network, so that research and extension activities can be geared to the needs of this group. This lesson is particularly relevant for information networks which, while providing a valuable service, do not appear, from evaluation data, to be adequately considering who will be using their services, and what information needs to be provided. For example, the evaluation of the IDRC Information Sciences Division (ISD) notes that:

> ...of the potential user community of researchers, agricultural planners, extension workers, librarians, and small- and large-scale farmers, the actual user group has rarely included any of the 40,000 small farmers who comprised the main target audience (Durrant, 1989).

Capacity building

Training, both formal and informal, has been a significant network activity, leading to increased confidence and abilities in areas such as undertaking research, report writing and designing projects. Five of the network evaluations considered training to have had

Table 2 Targets of information dissemination by IDRC-supported networks

Network	Researchers/ consultants	Governments	Other institutions	Ultimate beneficiaries
Research networks				
Adolescent fertility	Yes		Yes	Yes
ESARRN	Yes		Yes	
LAAN	Yes	Yes	Yes	
LAUHN	Yes		Yes	
MAP	Yes			
NAGA				
ONEASA			Yes	Yes
PFIAU	Yes	Yes	Yes	Yes
RISPAL			Yes	
Information systems networks				
AIBA	Yes			
INFOPLAN				Yes
IDRC-ISD	Yes	Yes		Yes
Technology transfer networks				
CTCS	Yes			Yes
RETAIN	Yes			
TECHNONET Asia				Yes

a major impact on the success of the network. However, more attention needs to be paid to making training appropriate to participants' needs.

Networks which provide inputs not locally available have strengthened institutional research capacity, and have led to the establishment of national or regional institutions or fora which did not exist prior to the formation of the network. Among the examples of this are the Agricultural Information Society of Asia, which was created as a result of AIBA, and the Asian Industrial Extension Officers' Forum, formed under TA. In another case the Macro-economic Analysis Program (MAP) led to the revival of the *Eastern Africa Economic Review*, which 'provides a badly needed forum for debating on economic policy issues relevant to the region, disseminating original research results, and linking economists in the region' (Young and Wangwe, 1985). AIBA also made a significant contribution to the formation of the International Information System for the Agricultural Sciences and Technology (AGRIS). Through the IDRC-ISD Caribbean Program, the Jamaican information system has served as a model for the establishment of similar systems in Barbados and the Dominican Republic. Information systems have also been established in Chile and Argentina following initiatives taken by the Latin American Regional Information System on Planning (INFOPLAN). However, the experience of six of the networks suggests that institutional capacity is sometimes not strengthened uniformly across the network, and that care must be taken that one participating institution does not benefit at the expense of others.

Cost-effectiveness

Networking has, in most cases, proved to be a more cost-effective method of delivering support to national programmes than the alternatives. The evaluation of the Caribbean Technology Consultation Services (CTCS) compared the cost of delivering network services with those of the United Nations Industrial Development Organization (UNIDO), and found that CTCS costs were 50% lower (Stanley and Elwela, 1988). Likewise, the evaluators of LAAN costed various research alternatives and found that it was cheaper to fund a network than to fund individual research projects (Moreau, 1991).

Summary of Lessons

Networks must evolve in relation to the needs of their members. If they ignore this cardinal rule they run the risk of becoming outmoded or dysfunctional. Networks need to monitor their operations and strategies to ensure continuing relevance and success. This includes ongoing or periodic assessment of all networking activities, leadership and coordination roles, network services, communication among members, dissemination of information, and adequacy of reporting, monitoring and evaluation procedures. Assessments in which the network actively participates tend to have more influence over subsequent network activities.

Planning for post-donor viability is an essential part of network planning. The literature suggests that donor support to a network should be defined from the outset, in terms of both nature and duration. If the duration of support is not discussed, and the question of the network's sustainability following termination of support is not explicitly addressed, the planning of the network's activities will be unlikely to take this issue adequately into account and the search for alternative sources of funding will be ignored.

By encouraging cooperation among research institutions and demonstrating positive results, networks can facilitate and encourage political commitment to a strong national programme (Moreau, 1991). National capacity is also strengthened through the development of a critical mass of national researchers and through access to regional expertise.

Networks must take greater care in determining who are their beneficiaries and in gearing their services to that group. Too often, network services remain unknown to those who could benefit most from them.

Conclusions

Monitoring and evaluation are critical to addressing issues related to the everyday management of networks, the quality of the training and services they offer, and hence to the general satisfaction of members. However, based on the literature reviewed and on IDRC's experiences, both approaches have been less than precise in assessing such factors as network impact at the national level, long-term viability and operational performance. Another area where we need to sharpen our evaluation tools and skills is in assessing the relative efficiency of collaborative research through networks compared with researchers working independently. With the recent proliferation of research networks, hard information on this issue will be increasingly important both to donors and to the institutions concerned. We therefore offer some suggestions for strengthening the application of both internal and external evaluations.

Data from internal evaluations are usually timely but tend to be impressionistic or anecdotal. They are not usually based on rigorous methods of data collection, tending instead to draw on the perceptions of those closely involved in the ongoing operation of the network. By their very nature, internal evaluations lack a broader strategic perspective and cannot deal well with sensitive issues. External evaluation, on the other hand, can be much more rigorous in its data collection methods and can yield quantitative data. The analysis of these data from an external perspective can take strategic considerations and contextual and external factors into account. Contentious issues can be more easily dealt with by taking a broader perspective. The usual problem with external evaluation is that some data must be collected retrospectively and may therefore be of poor quality.

We submit that both internal and external evaluations are useful. In fact, the two should buttress each other. Monitoring will continue to be part of the day-to-day management of a network, while external evaluations will continue to be required for accountability or to document performance. Each approach has strengths not found in the other. IDRC is currently relying more and more on an approach which combines both, with internal evaluation informing external evaluation and vice versa. To make internal evaluation more effective, more attention should be paid to data collection methods; and more elements of external evaluations should be subsequently built into network activities.

Both internal and external evaluations can contribute greatly to levels of awareness and cooperation among network members. To be effective, both approaches should be highly interactive and should build toward systematic and ongoing review of the daily life of the network. Results should permit scientists, policy-makers, donors and others to be aware of emerging problems and to make changes early on rather than wait until problems become ingrained. Evaluation results should be regularly disseminated among members through newsletters or other communications, to expose them to the value of this kind of exercise and the insights available through it.

The results of evaluations which are participatory in nature are more readily accepted and implemented. All those directly involved in the network should have a say in the design and implementation of evaluations. There should be no monitoring or evaluation mystique: participants should feel free to design the data collection methods and to specify the issues appropriate to their concerns and to the nature of their network. Wherever appropriate, evaluation activities should be integrated in existing data collection, review and reporting mechanisms, and should not impose a significant new burden on network participants. Ideally, participants will identify with the process and be prepared both to contribute to it and to benefit from it.

References

Akhtar, S. 1990. Regional information networks: Some lessons from Latin America. *Information Development* 6 (1): 35-42.

Banta, G. 1982. The use of networks to strengthen the crops and cropping systems group activity: A discussion paper. IDRC, Ottawa, Canada.

Buzeta, R. 1989. Aquaculture development thrust (regional). IDRC project summary 89-0017. IDRC, Ottawa, Canada.

Durrant, F. 1989. Evaluation of the Caribbean Program, IDRC Information Sciences Division, 1970-1987. IDRC, Ottawa, Canada.

Faris, D. G. and Ker, A. (eds) 1988. Eastern and Southern Africa: Network coordinators' review: Proceedings of a workshop held at Nairobi, Kenya, 9-12 May 1988. IDRC, Ottawa, Canada.

Faris, D.G. 1991. Agricultural research networks as development tools: Views of a network coordinator. IDRC, Ottawa, Canada, and International Crops Research Institute for the Semi-Arid Tropics, Patancheru, Andhra Pradesh, India.

Gastal, E. 1987. Cooperative activity and efficiency in agricultural research. In: Webster, B., Valverde, C. and Fletcher, A. (eds), The impact of research on national agricultural development. International Service for National Agricultural Research, The Hague, Netherlands.

Jarmai, L.J. 1982. TECHNONET Asia: An evaluation. IDRC, Ottawa, Canada.

Ker, A.D. 1989. Rootcrop Network (Eastern and Southern Africa): Project summary submitted to IDRC Board. IDRC, Ottawa, Canada.

Koala, S. and Banta, G. 1989. West African Farming Systems Research Network, phase 2: Project summary submitted to IDRC Board. IDRC, Ottawa, Canada.

Moreau, L. 1991. Evaluation of the Latin American Aquaculture Network. IDRC, Ottawa, Canada.
Plucknett, D.L., Smith, N.J.H. and Ozgediz, S. 1990. *Networking in international agricultural research*. Cornell University Press, Ithaca and London.
SPAAR. 1987. Collaborative research networks: Desirable characteristics. SPAAR, Washington, D.C., USA.
Stanley, J. L. and Elwela, S.S.B. 1988. Evaluation report of the Caribbean Technology Consultancy Services (CTCS), CTCS Network Project (1985–1988). IDRC, Ottawa, Canada.
Valverde, C. 1988. Agricultural research networking: Development and evaluation. International Service for National Agricultural Research, The Hague, Netherlands. Staff notes (18–26 November 1988).
World Bank, 1987. Regional cooperation in agricultural research. In: West Africa agricultural research review, 1985–86. West Africa Projects Department, World Bank, Washington, D.C., USA.
Young, R. and Wangwe, S.M. 1985. Macro-economic Analysis Programme for Eastern and Southern Africa. IDRC, Ottawa, Canada.

Address
Terry Smutylo and Saidou Koala, IDRC, P.O. Box 8500, Ottawa, Canada K1G 3H9.

Support to Networking in Africa and Latin America: The Role of AGRECOL

Irene Täuber, Almut Hahn and Claudia Heid

Introduction

In this paper we will describe the role of the Centre for Information and Networking for Ecological Agriculture (AGRECOL) in encouraging networking among the sustainable agriculture movements in Africa and Latin America. We should begin by emphasizing that AGRECOL is not a network itself but a Northern partner of African and Latin American networks. The role of AGRECOL is that of a catalyst and a bridge, linking the sustainable agriculture movements of the South with the information and financing organizations of the North. Being continuously in contact, we are able to serve as a partner in dialogue to both sides, and to support the networks of the South with specific expertise when needed. Our tasks are continually redefined by the growth of our partners and their changing needs. As an information and documentation centre in the North, we are concerned with the regionalization of information services and the documentation and synthesis of experiences and knowledge.

How did we come to focus on networking as one of our main activities? We will first briefly look back at AGRECOL's origins and explain how it has come about that we do now what we never planned to do at the beginning! We would still like to get a clearer idea of what support services can sensibly be implemented from the North. We are therefore interested in strengthening our dialogue not only with partners in the South but with similar institutions in the North as well.

Evolution of AGRECOL

Origins

In 1982, when the Green Revolution and its so-called benefits were being heavily promoted, not much was known about experiences in environmentally sound agriculture among people concerned with development aid. However, such experiences did exist, and more and more people were looking for environmentally sound solutions to small-scale farming problems. Many of these were lone fighters.

At that time several like-minded individuals from different disciplines working in development in Germany and Switzerland came together. They decided to form a working group and founded the AGRECOL information centre, backed by an advisory board.

In 1983, the concept of the AGRECOL project was still rather vague. The main objectives were to:
- Create a centre where information is collected and retransmitted.
- Collect documents on existing experience in sustainable agriculture in the tropics and sub-tropics.
- Facilitate contacts to improve the exchange of experiences.

The information centre was to be located in the North. It remains a fact that a great deal of information still comes together in the North. In addition, South-North-South communication channels usually work much better than South-South ones. It is on the whole easier for Northern institutions to obtain information from universities and agricultural research centres.

Initial activities

We began by seeking to set up a library, at our headquarters in Switzerland, on sustainable agriculture in the tropics and sub-tropics. We felt that specialized information on this topic was greatly needed and would complement the documentation services of development agencies, universities and others. The library contains books, magazines, reviews, and a lot of grey literature—information which is not available through bookshops, such as technical leaflets, brochures and training materials. These documents are mostly compiled by technicians and advisers.

We also set up a data base of expertise by asking people to write down their fields of experience and to give us contact adresses of more people and organizations.

We then announced the existence of AGRECOL by informing various magazines and newsletters, asking them to write about us.

Questions and answers

In response to their articles about AGRECOL, we received many letters, especially from interested field workers and colleagues in Southern NGOs. We then started a question-and-answer service. We received questions of a general nature on sustainable agriculture, as well as very specific technical questions on plant protection, erosion control, seed conservation, training opportunities, and so on. Books and photocopies of materials available in our library were sent on request. Our preferred course of action, however, was to link these requests for information to resource persons or organizations working in the same region or country as where the question came from.

AGRECOL thus began functioning as a turntable, linking information and people in a South-South as well as a South-North context. European development volunteers visited us as part of their preparations for departure, but so also did representatives from Southern non-government organizations. To both we offered individual consultancy services and user-friendly documentation.

Relationships develop

Out of these contacts and visits, deeper partnerships gradually developed. We came to know people and organizations with goals similar to ours. In dialogue with these partners and through their feedback, we learned what information is really relevant for our target group, and in what form it should be disseminated.

We became convinced that information is most sucessfully exchanged on a local basis. Local information services would better meet the specific information needs of a region or country, to which it would be culturally adapted in content as well as in form. Rapid, low-cost access to tailor-made information would be easier to achieve through local information centres. So we decided to begin supporting the development of information services in the South.

In addition, we began to realize that written information is one way, but not the only nor even the most appropriate way, of spreading knowledge in agriculture. To organizations in the South, oral communication and practical demonstrations are very often more important. This results in the need for more pervasive exchanges, involving materials, visits and meetings as well as written documents, with people working with similar approaches. These exchanges are best facilitated through networks. Supporting networks therefore became one of our principal activities.

AGRECOL further decided to concentrate on restricted geographical areas: Africa, with an emphasis on West Africa, and Latin America, with an emphasis on the Andean countries. To get a network going you need intense discussions and detailed involvement with your partners. Only by getting to know each other well can you create confidence and start to share tasks in a way that increases effectiveness while saving on resources. With limited staff, this is only possible if you limit the geographical scope of your activities.

This strengthened relationship with Southern partners and greater involvement with networking led to our being invited to assist in the organization of the 1989 conference of the International Federation of Organic Agriculture Movements (IFOAM). This was held for the first time in a developing country, in Ouagadougou, Burkina Faso. By raising funds for scholarships, many NGO partners from developing countries were able to participate. Numerous initiatives in working together were made as potential partners met face to face for the first time.

Supporting Networks

Regional initiatives

Following the IFOAM conference of 1989 we entered a period of reflection. Many of our partners had expressed a wish to organize regional meetings which would serve their needs more specifically than a world conference. They needed to tackle concrete problems in their daily work through exchanges on a local basis.

In October that same year, 60 people from all over Latin America came together in a seminar convened by IFOAM in Cochabamba, Bolivia. All the participants, mostly NGOs, were working on sustainable agriculture and were interested in improving exchanges through networking. As in Ouagadougou, AGRECOL co-organized this seminar. At this meeting the participants defined a structure of five Latin American sub-regional networks supported by supra-regional thematic working groups on subjects such as information, training, research and marketing (including certification of organic produce).

At this stage AGRECOL redefined its tasks as follows:
- Facilitate contact between interested persons
- Take part in early network activities, such as initiating workshops
- Provide practical support for preparing and realizing meetings
- Link partners with potential donors
- Participate in network meetings
- Attend to the follow-up of these meetings, where appropriate.

Networks Develop

Africa
We had imagined that out of the 50 West African participants at the Ouagadougou conference an active working group would develop. We had invited them one evening to an extra meeting. All of them had expressed the keen desire to stay in contact and to form a network. One organization in Senegal was asked to be the contact address; everybody should write to them. But, we were told, in 8 months only two letters from Africans had arrived. Thus we learned that a network cannot be created by mail, at least not in the oral culture of Africa.

Discussions with partners led to the idea that the exchange of experiences should first take place at national level, and that NGOs, government services and farmers' associations should be included.

We had contacts with many people, all of whom said that more exchange was important. AGRECOL therefore took an initiative at the IFOAM Conference in August 1990. There we talked with three representatives of Senegalese organizations and promised support if they wanted to build a network. They formed a committee, we gave them the contact addresses we had. On their return to Senegal they invited more than 100 people to a meeting to launch the network, 35 of whom attended. AGRECOL helped to find funds for this meeting, which was held in June 1991. Finding money for a project which did not yet exist was difficult. Donors do not like to spend money on discussing plans when it is not clear what the plans will look like. The organizing committee therefore reduced the duration of this first meeting to 2 days, to make it cheaper and thus easier to fund. During these 2 days everyone had an opportunity briefly to present his or her experience. After much discussion on whether the structure should be formal or informal, a 'light' structure was decided upon, with national coordinators and five regional coordinators. An inventory of expertise of the new network's members was compiled and distributed. It consisted of a list of members' addresses, the experience and resources they could offer, and the topics on which they were looking for information.

After this promising start, difficulties began to surface. Lack of finance and misunderstandings hampered the organization of regional meetings. Some coordinators have met, but one is overloaded with other work, another has gone abroad for training, and still others have left their NGO and have been replaced by new staff who did not attend the original meeting. The 'grass roots' would like to do something, but they await leadership from 'above'. There are now plans to organize a general meeting to discuss the statutes of the network, to choose the new coordinators and to decide on a working programme.

Other iniatives have been somewhat more successful. With or without the support of organizations such as AGRECOL and ILEIA, networks are emerging in many other West African countries, such as Burkina Faso, Ghana, Benin and Cameroon, to mention but a few.

Latin America

Since the Cochabamba meeting of 1989, networking at national level in Latin America appears to have become better established. AGRECOL is now in close contact with national networks in Mexico, Peru, Ecuador and, recently, Colombia. In some cases these initiatives are supported by a strong grass roots movement. In others, such as Bolivia, Chile, Paraguay, Argentina and Brazil, people are starting to meet and to discuss possible ways of cooperating. These national networks are very diverse, varying according to national conditions and needs. But all of them have a strong focus on information exchange, training opportunities and advisory services. Participants frequently express the need for good communication with relevant organizations in the North.

Supra-regional networking depends greatly on the capacities and commitment of individual sub-regional coordinators. The supra-regional thematic working groups planned at the Cochabamba meeting have not been successful. However, information is flowing well within specific regions and countries, and between these and some Northern partners.

At the second regional meeting, held in São Paulo in November 1992, about 60 participants met for 4 days. Papers had been prepared in advance on all the topics to be discussed. Reports were presented from the different regions and from national networks. There were also reports on workshops on priority issues for the movement (such as training, information, certification and statutes). The difference between Cochabamba and São Paulo was that, for this second meeting, everybody worked on the basis of her/his concrete experience and everyday needs. Targets were set more pragmatically. Programmes had to be designed strictly in accordance with the network's priorities and its ability to implement them. The mixture of about 40% 'old boys (and girls)' and 60% 'newcomers' seemed especially beneficial. AGRECOL participated in this meeting as a Northern partner of the network.

Achievements

What has been achieved, and how? AGRECOL's question-and-answer service is an instrument for informal networking. Its effects are difficult to predict: many seeds are sown, only a few will grow. But we believe that through it more people have come to know about the principles and techniques of sustainable agriculture, to implement them, experiment with them and improve or adapt them to local conditions. The service is often the subject of discussion. It is quite time-consuming and we would like to have it more formally organized, with its own budget. Nevertheless, it is an important service and a useful tool for keeping in contact with a broader public.

In addition, partners with similar approaches and problems have got to know each other and now communicate. As an independent NGO in the North, being neither church nor government nor donor, we have a neutral position and can bring people together who would not otherwise meet. Our independent stance is greatly appreciated, and often helps us play a catalytic role.

At national level, a dynamic networking process is now in place. Judging by what we see, just in a Northern information centre such as AGRECOL, a great deal of information is now being circulated in the South.

We in the North act as a relay station, making a lot of information available to partners in the South. We are also mediators, a bridge, a turntable for information between South and North. As the partner of Southern networks dedicated to sustainable agriculture, we contribute to raising consciousness in the North of the importance of sustainability issues in developing countries. Through our centre, volunteers have an opportunity to prepare themselves for their work in the South.

We have published a number of bibliographies, workshop papers and compilations on special topics, and hope that these are useful.

Lessons Learned

What can we learn from all this? This section outlines some of the lessons.

Networking
First of all networking is attractive only for short periods, that is during a conference! Potential members of networks can imagine the benefits, such as improved information

flows, better division of labour, or more training opportunities, but the efforts needed to realize those benefits are frequently underestimated round the conference table, only to seem too daunting once the participant returns home to the more sobering realities of his or her day-to-day working environment.

Networking always requires inputs: letters have to be written, seminars organized, contents of a meeting planned. One's own experience has to be documented if it is to be shared. While networks can subsist on goodwill in the short term, in the long run these inputs are not possible without some financial assistance, although the amount need not be large.

Northern support organizations should not forget to support the follow-up process required after meetings. Like their Southern counterparts, however, they should recognize the limits of their capacities and not promise more than they can deliver. For us at AGRECOL, this means that we should limit our geographical focus, at least during the start-up phase.

We should also concentrate on deepening our understanding of familiar subjects rather than on broadening our subject matter coverage. In AGRECOL's case this means focussing on advising national or regional information and documentation services and on providing organizational support.

To start a network successfully it is essential to build up confidence and to foster strong links between potential collaborators. A short workshop—only 2 days due to lack of funding—was clearly not enough in the African context. A committed person or, better, a group acting as 'animator' can, however, stimulate the development of a network.

The roles of coordinators and grass roots members must be made quite clear. The relationship between them should not come across as vertical. Especially in Africa, the temptation to resort to hierarchical relationships seems great.

As well as a desire to collaborate there may be competition between organizations, although each may be sincerely committed to sustainable agriculture. This should not be ignored, but analyzed and handled in a healthy way.

Networking is an organic process of growth which cannot be determined by blueprints and linear planning. We must be clear about the justification for a network and about its general aims, but its activities must remain flexible. There must be sufficient time to get to know each other and to gain mutual confidence. Nevertheless, network partners need to benefit quickly from their efforts to launch a network.

A network consisting not of protagonists solely from the same background (e.g. all farmers) but integrating different levels (farmers, researchers, consultants, etc) must define its aims, its methods and its language according to the needs of the weakest partners. The initial steps in a networking process must be small and not too demanding. Responsibilities must be realistically described and delegated. It is important that every member taking on a task in the network realizes how much time or other resources will be needed to successfully carry out that task.

Network members have different backgrounds...

Information and documentation services

The regionalization of information services is an important step in promoting sustainable agriculture. Practical information should be locally available to facilitate quick and easy access. Local information services are better able to judge the relevance of materials to users than are international ones. A local or national information centre can also serve to document and synthesize experiences close to the grass roots level. Such a centre differs from a Northern international information centre such as AGRECOL, in that it can keep its finger on the pulse of initiatives and experiences in the field. This leads easily and naturally into a networking role.

Many institutions have no resources for setting up a documentation centre and for collecting, processing and disseminating information. They need useful information materials, training, advice and access to experiences in their own and other regions. This is why AGRECOL supports local initiatives in launching information and documentation services, using its own experience in setting up an information and networking centre as a basic resource. In turn, we consult our partners in planning and implementing such projects, asking them to help select and recommend useful material.

In Latin America we had an experience in supporting the launching of a local information centre that showed us the limitations of a consultancy based on an informal relationship instead of an official contract. This information service developed in a way that was completely alien to our concept and intentions. The service simply does not function to meet the needs of local organizations and people working on sustainable agriculture, as it was planned to do. The NGO hosting the service seems not to recognize the potential advantages of sharing. Instead, it sees information management as another way of controlling resources and increasing its power. It does little to support the independent development of the service, keeping it small and introspective, and hindering its networking activities.

This experience taught us that an informal consultancy can be fruitful only if both sides share the same concept and intentions. Next time, more discussions beforehand, better monitoring and a clearer definition of the project and its tasks will be needed.

Evaluation

Evaluation is vital to maintain the health and relevance of network activities in a rapidly changing world. Our own internal evaluation, carried out in 1991 with the assistance of an external consultant, proved extremely useful. It involved regular, intense discussions with our project committee. We also sent a questionnaire to our partners in Africa and Latin America. Other useful monitoring and evaluation tools include detailed annual and monthly planning meetings, days of retreat by the team for special subjects and decisions, visits and discussions with our partners in Africa and Latin America, and professional exchanges with Northern partners.

Address

Irene Täuber, Claudia Heid and Almut Hahn, AGRECOL, c/o Ökozentrum, Langenbruck, CH 4438 Switzerland.

Supporting Regional Networks: The Experiences of ILEIA

Wim Hiemstra and Carine Alders

Introduction

A survey conducted in 1981-82 among Dutch development workers confirmed the need among field-based development organizations for information on agricultural development in situations where farmers have little or no access to commercial inputs such as chemical fertilizers, pesticides and machines. A decade later, many farmers still have limited access to such inputs, while awareness of the disadvantages of using chemical inputs is growing. The survey revealed that many fragmented activities related to low-external-input and sustainable agriculture were conducted at grass roots level, with little or no systematic guidance from established research and training institutions. Nor was this multitude of experiences being well documented, as conventional publications focused mainly on conventional, high-input approaches. In an effort to remedy this situation the Information Center on Low-external-input Agriculture (ILEIA) was created with the aim of establishing a modest documentation unit, an information service and a quarterly newsletter. The project was funded by the Dutch Ministry of Development Cooperation.

The objectives of ILEIA are:
- To collect information on low-external-input agriculture.
- To disseminate this information, mainly to field workers working with small farmers.
- To make development agencies (researchers, educators and policy makers) aware of low-external-input approaches to agricultural development and to promote the dissemination of such approaches.

To reflect the growing concern over the sustainability of agriculture and the links between sustainability and the low use of external inputs, it was decided soon after foundation to change the name of the centre, which became the Information Centre for Low-external-input and Sustainable Agriculture. (The original acronym was, however, retained, because of its familiarity.)

The Network Develops

A global start
This information exchange network started out with considerable emphasis on the centre as the collector and disseminator. Tools for exchanging information included the

quarterly *ILEIA Newsletter* and a modest question-and-answer service, both fed by a documentation unit. Workshops were a third tool for gathering, analyzing and disseminating information. ILEIA staff asked people to write down their experiences in articles for the newsletter and to search for and send in information to be documented in the library. Without special efforts from ILEIA's side, the number of subscribers to the newsletter grew rapidly. Although the main target group is still field development workers, many trainers, researchers, policy-makers and others interested in low-external-input agriculture have been added to the mailing list. This means that our sources of information have also grown in number and diversity, since the newsletter is also the means by which we ask people to contribute their experiences.

During the early phases of the project, the global perspective brought both advantages and disadvantages. As limited information was available, it had to be gathered on a global scale to get the network going. Pioneers in low-external-input agriculture felt that reading about experiences from around the world was encouraging and inspiring. On the other hand, such agriculture is, by its very nature, location-specific, so there were (and still are) articles in the newsletter that were not relevant for certain readers, because their specific farming system was different.

The number of readers of the *ILEIA Newsletter* has continued to grow rapidly in recent years (Table 1). The newsletter has thus proved a highly effective tool for building the ILEIA network. It is a means of contact, it invites people to join and participate. Adopting a thematic approach, it spontaneously triggered new networks with each issue.

Table 1 Growth in readership of the *ILEIA Newsletter*, 1988-93

Readership	31/12/88	01/06/91	31/12/92	01/06/93
Total	2500	4412	5666	6800
Developing countries	1820	3316	4189	5500

Newsletter as a networking tool

The *ILEIA Newsletter* has always been our main networking tool. The extent to which it triggers horizontal communication (readers contacting each other) was one of the questions asked in the readers' survey conducted in 1992. It was found that the main readership groups are field workers (25%), trainers (25%) and researchers (27%). Government and non-government organizations (NGOs) appeared to be about equally represented. Articles in the newsletter described experiences from many different points of view, including that of the farmer as well as those of trainers and researchers. The newsletter is thus a passive link between these different groups of readers and writers. The same passive networking link is established between readers from the North (14%) and

South (87%). Some attention is given to experiences in the North to keep readers from the South abreast of developments.

The language used in the newsletter aims primarily at middle-level field workers. This seemed to be no problem for other target groups. Most respondents (84%) judged the level to be appropriate.

Half the respondents judged the newsletter to be useful in facilitating networking. Trainers were somewhat more positive than were field workers and researchers. The *ILEIA Newsletter* is often shared with other people. The readers' survey showed that 93% of subscribers share their newsletter with at least one other person (Table 2).

Table 2 Number of other people sharing the copy of the *ILEIA Newsletter* received by the respondent

Number of other people	% of respondents
0	6.5
1-5	48.1
5-10	25.4
10+	19.9

The newsletter also functions as an information broker. More than half the respondents had asked for further information on at least one of the publications mentioned or reviewed in the newsletter. More than 18% of them had asked for references, publications or further information five or more times in 1 year (Table 3). To facilitate direct contact, ILEIA always publishes addresses of authors and publishers whenever possible.

With regard to the exchange of information on low-external-input agriculture, 80% of respondents said that ILEIA had been instrumental in this, although the answers varied from 'a little' to 'absolutely'. The ways in which ILEIA has been of use also differ. Some respondents talked about the strengthening of their ideas, or of the obvious new flow of information represented by the newsletter. Others pointed to the translation of articles into Spanish or other languages. Translation is a common means of further disseminating information in the newsletter. In its newsletter *Nouvelles de Pronat*, the NGO Environnement et Développement du Tiers Monde (ENDA) regularly translates *ILEIA Newsletter* articles into French. The Brazilian organization Assessoria e Serviços a Projetos em Agricultural Alternativa (AS-PTA) translates *ILEIA Newsletter* articles into Portuguese for its newsletter *TA em Periódicos*. Translations into Hindi, Oriya, Tamil and

Table 3 Number of requests for further information arising from the *ILEIA Newsletter*

Number of requests per year	% of respondents
0	26.3
1-5	55.3
5-10	13.2
10+	5.2

Thai have also been reported. A social forestry newsletter in India publishes information drawn from the *ILEIA Newsletter* and reprocessed to suit the needs of its readership. All information published in the *ILEIA Newsletter* is free of copyright restrictions and readers are encouraged to make as many photocopies as they need. 'Photocopyability' is a criterion used in designing the layout of the newsletter.

Attempts at regionalization
As more and more people become interested in low-external-input and sustainable agriculture, more relevant information becomes available and more site-specific questions are asked. The number of subscribers to the ILEIA network has now grown to such an extent that a global register of members is no longer feasible. In 1993 we therefore intend to publish the register by continent. This issue made us reconsider ILEIA's role. We decided that besides functioning as a global network ILEIA should support local and regional information networks, allowing more intensive, site-specific exchanges of ideas and information.

Our policy has been to respond to individual requests for assistance in establishing regional newsletters and documentation centres, rather than to take such initiatives ourselves. But how best to stimulate and support regional networking? Where to start? Several tools and approaches were developed and tried out:
• In 1990, through a questionnaire sent to all subscribers, we had made a survey of regional NGOs and individuals with an interest in sustainable agriculture, including their goals and activities. This information was sorted by country and published in a Register (a special issue of the *ILEIA Newsletter*, namely vol. 6 No. 4).
• ILEIA staff have actively supported local initiatives in launching networks by assisting in the organization of founding workshops and by supplying seed money for surveys of potential members and the writing of project proposals for submission to donors.
• A small libraries programme was started as a result of the AGRECOL/ILEIA bibliography *Towards Sustainable Agriculture*, published in May 1988. Readers of the *ILEIA*

Newsletter intending to start a small local library could order their selection of books to a maximum value of Dfl. 1000 per request. A total of 140 requests were received. ILEIA sought and found funds for delivering the books. After reviewing this whole exercise within the European Network for Low-external-input and Sustainable Agriculture (EULEISA) (a network of European support organizations), a second phase of the small libraries programme is being prepared. In future, organizations that have received a small library may be approached to cooperate in a regional network.

• ILEIA has also attempted to stimulate regional networks by asking members to function as regional contact persons. The idea was explained in a newsletter article and people were asked to react. These people should be willing to exchange information, for example by writing articles on field experiences with low-external-input and sustainable agriculture in their region and/or network. ILEIA would facilitate this exchange of information. Contacts with 30 persons from 20 different countries were established. Mailings were sent round, consisting of contributions from the regional contact persons. In January 1991, ILEIA evaluated this experiment. It turned out that exchange on this more personal basis is very useful, but very time-consuming as well. We could not free enough time to make this method of exchanging ideas and experiences a success. We decided not to continue this method of networking and to concentrate instead on supporting initiatives from the regions on a broader institutional basis.

Support to regional networking
Several activities in support of regional networking were identified:
• Strengthening regional facilities for documenting, publishing and disseminating relevant experiences. This is accomplished through such activities as establishing or supporting existing national or regional libraries and/or documentation centres devoted to low-external-input and sustainable agriculture, establishing or supporting regional newsletters on low-external-input and sustainable agriculture, and recording relevant field experiences through written case descriptions and visual productions.
• Strengthening regional cooperation and exchange of information through support to regional initiatives. This includes the organization of regional seminars on low-external-input and sustainable agriculture, assessing the status of relevant agricultural practices in the agro-ecological zone(s) of a region and formulating programmes for their further development, and legitimizing low-external-input and sustainable agriculture approaches among regional field staff and national and international policy-makers.
• Strengthening South-South cooperation and South-South and South-North exchanges of information. Activities here include supporting the establishment of networks, linking networks within and between regions as well as networks from different language areas, pooling human resources for mutual support activities through the exchange of staff (South-South and South-North) and by facilitating South-South consultancies and training events, and the publication of reports of activities and experiences, either in the *ILEIA Newsletter* or in the different regional newsletters.

Supporters work together

In Europe, ILEIA participates in a network of like-minded organizations, including the Swiss-based Centre for Information and Networking for Ecological Agriculture (AGRECOL) (see Täuber et al, p.249). These organizations meet twice a year to exchange experiences, inform each other of their activities and discuss common action. For instance, a common thesaurus on low-external-input and sustainable agiculture has been developed to make our documentation systems compatible. A database on training opportunities is now being developed, together with an inventory of donors interested in supporting regional activities. The costs of this European network are met by its participants. Secretarial tasks such as writing the minutes of meetings rotate informally. Often participants contact each other on topics that may not interest the whole network.

Regional Networks Emerge

At the request of organizations in India, the Philippines, Benin, Sri Lanka and Ghana, ILEIA staff members visited these countries between 1989 and 1992. Various networking activities developed, in which we played different roles. In 1990 we evaluated our experiences, with the aim of assessing progress and drawing out the major lessons. The comments below represent our own assessment, and may not completely reflect the opinions of our regional partners.

The Philippines

In the Philippines, ILEIA took part in a workshop organized by the International Institute of Rural Reconstruction (IIRR) and the Agriculture, Man and Ecology (AME) programme, to which former participants in AME's courses were invited. Participants at the workshop decided to set up an association of professionals committed to low-external-input and sustainable agriculture and to review the options for creating a national network/ information centre. A steering committee was formed to facilitate cooperation and a person was contracted, with financial assistance from ILEIA, to compile a register of interested people and institutions. An initial survey of experiences and expertise yielded an unexpectedly high response and revealed a great number of activities and documents. The steering committee organized a second workshop, this time inviting other interested people. The outcome of this workshop was the formation of a committee to prepare a proposal for a 4-year programme and the formation of an editorial committee for a newsletter.

The networking process was positively influenced by the fact that AME alumni from the Philippines had a common background (the AME courses) and by the availability of funds for creating a register, the second workshop, and the first issue of the newsletter. However, some constraints were also encountered. There was uncertainty regarding the need for yet another national level organization. NGOs engaged in protracted discussions on the most desirable level (national or regional) and focus of the network. The complex infrastructure in a country of many islands did not make matters easier. Although no formal network came out of this initiative, the register proved valuable for those interested in low-external-input and sustainable agriculture in the Philippines.

Recently, the initiative was taken to start a network on calcareous soils. ILEIA made seed money available to organize a first get-together.

India

Three organizations (a local NGO called Kudumbam, the AME project in Pondicherry, and an existing social forestry network in Tamil Nadu) initiated a network on low-external-input and sustainable agriculture in Tamil Nadu State (see Quintal and Ghandimathi, p.177). Two workshops were held in 1990. ILEIA staff took part in the second workshop, on networking. Some 30 NGOs decided to institutionalize the network, start a newsletter and establish a documentation centre. A funding proposal was formulated and funding was granted by the Dutch co-financing organization, Humanistisch Instituut voor Ontwikkelingssamenwerkung (HIVOS).

The start-up of this network was positively influenced by the fact that a network on social forestry already existed. Articles on regionalization in the *ILEIA Newsletter* prompted the initiators of this network to contact ILEIA for support. We responded by making seed money available for two workshops and a newsletter and by participating in the second workshop. A slow decision-making process on the part of the donor slowed down the start of the network, however.

In 1990 ILEIA was asked by the nascent Gorakhpur network in Utar Pradesh State to support the translation of *ILEIA Newsletter* articles into Hindi. ILEIA agreed to look for donors for this task, and sent a representative to visit the network to help prepare a project proposal. The proposal included the production of a local language newsletter and the organization of several local workshops. Funding was granted by another Dutch co-financing organization, the Netherlands Organization for International Development Cooperation (NOVIB). A list of several hundred potential subscribers was compiled as an informal register.

In this case the *ILEIA Newsletter* triggered the establishment of a new local network. Having an ILEIA representative available in India to discuss ideas contributed greatly to the process. ILEIA made seed money available for the first year of the 3-year proposal.

Ghana

When the agricultural coordinator of the Ghana National Catholic Secretariat attended a workshop of the Northern Ghanaian Association of Church Development Projects, he was inspired to start a network for the southern part of the country. This idea was introduced to an Oecumenical Committee, which formed the nucleus from which the network was developed. All kinds of different parties were invited to join, and the group immediately appointed an ad hoc committee to prepare the ground for establishing the network.

The costs of travel and logistical support for this ad hoc committee were met through seed money provided in the form of a loan from ILEIA. This enabled the committee to formulate a provisional statement of intent and to visit potential members to discuss the need for a network and the details of its establishment. This seed money also enabled the committee to seek funds from donors for organizing a workshop at which the structure, objectives, activities and membership of the network would be determined. After the workshop, the ad hoc committee handed over its responsibilities to a democratically elected Executive Committee and the network was officially launched.

Personal links between ILEIA staff and Ghanaian counterparts facilitated ILEIA's support to this network. The provision of seed money appears to have been critical. The network now has 34 members and a secretariat to take care of its daily affairs. It has organized study tours in West Africa, and conducted training courses in low-external-input and sustainable agriculture and participatory technology development. Its members are enthusiastic advocates of the further development of these approaches by national organizations.

Sri Lanka
The initiative for a network emerged in 1990 and the concept is still at a very early stage. Contact with ILEIA was established by a regional partner through meetings of the International Federation of Organic Agriculture Movements (IFOAM). Contacts within Sri Lanka were made through the newsletter and through ILEIA's regional contact person programme. An ILEIA staff member visited the country in late 1990 to explore the possibility of networking. The commitment of field organizations to the idea of building a network seemed to be limited. Since then no further steps have been taken by ILEIA.

South Pacific
A proposed network in the South Pacific Region also failed to materialize. One of the reasons could have been that the initiative came from university-based expatriates and did not appear to have broad-based support from national professionals. Another problem was difficult communication within the region because of its great distances. Moreover, nobody from ILEIA was close by, or had working experience in the region. We therefore lacked a good understanding of the situation, the organizations and the people concerned.

Benin
Having received the register of *ILEIA Newsletter* readers, a reader in Benin contacted other readers in the country to gauge their interest in a national network. The response was positive. A committee was formed and received seed money from ILEIA for further preparations. These included several workshops, field trips and the development of a

project proposal. In 1992 the Réseau de Développement d'Agriculture Durable (REDAD) was established. The Interkerkelijke Organisatie voor Ontwikkelingssamenwerking (ICCO), a third Dutch co-financing organization, has granted funds for a 3-year programme that includes activities such as publishing a newsletter (*Nouvelles de Redad*), achieving formal recognition, forming an information centre, providing assistance to farmers' organizations, and organizing general meetings and thematic seminars.

Lessons Learnt

Past efforts at regionalization
Several conclusions can be drawn from these attempts to encourage regional networks:
- Seed money is vital if a network is formally to come into being.
- Support at the international level requires a good understanding of the local situation, the institutions and the people. Both personal contacts and financial support appear necessary.
- Face-to-face contact between members is essential when the network is seeking to emerge. The geographical area covered by a network should be such that informal and/ or formal meetings are possible fairly frequently. This leads to more emphasis on small-scale sub-national rather than national, regional or international networks, depending, of course, on the size of the country and its communications infrastructure. Networks emerge most strongly when the initiative for starting them comes from grass roots organizations. Other organizations should be invited to join on the basis of their complementarity with these grass roots organizations. National and regional networks may start to emerge more strongly once sub-national networks are in place.
- It seems important that at the early stages the idea of launching a network should be shared by several people rather than just one person or organization.
- The process of establishing a network takes time—rather more than 1 year. The bottlenecks are communication up and down the institutional hierarchy, the workload of the people involved in the initiative and the time needed by donors to process project proposals.
- To prevent overlap and the duplication of efforts, it is important to inform partners within the European network of support organizations at an early stage of plans to support an emerging network.

ILEIA does not aim to be in the front line of support to emerging networks all over the world. It sees its role rather as a promoter of the idea, as a developer of the methodology to be followed and as a source of information for those who want to start a network. We should gather and disseminate information on how to establish networks. We also see a role for ourselves in developing a better understanding of networking—how it works and what it can add to development efforts. For this reason, a workshop on this subject was organized in 1992 in the Philippines, together with the International Institute of Rural Reconstruction (IIRR) and World Neighbors (see Haverkort et al, p.3, and the *ILEIA Newsletter* 8 (2)).

Looking ahead: What is ideal support?

Supporting regional networking initiatives has been a trial-and-error experience. We have learnt many lessons but still face dilemmas. Fundamental characteristics of low-external-input and sustainable agriculture include its diversity, complexity, site-specificity and long-term nature. Consequently, the requirements of support differ greatly from place to place, which means that there is no blueprint for supporting regional activities. The concept of a region varies too—from a group of countries to just part of one country. Below, we discuss some of the issues we have encountered in our efforts.

The first issue is the role to be played by ILEIA as the 'network centre'. Thus far our role has been limited to the backseat one of facilitating the formulation of needs, stimulating regional cooperation and supporting the formulation of work programmes and project proposals. ILEIA has acted as mediator in requesting funds. This has meant that regionalization has sometimes occurred more slowly than expected. However, when our role becomes more pushy, the fledgling network runs the risk of collapsing due to the limited motivation of members. Regional activities are therefore carried out only at the request of regional partners. It is unclear what our role should be once networks are established. This is an important issue for discussion at regional meetings, especially with individual network members. Should a minimalist approach in the use of Northern-based experts be pursued? What more can be done to enhance South-South cooperation? There still seems to be scope for ILEIA to perform a role as go-between, building bridges among organizations and networks. One disadvantage of assuming a backseat role is that competition among NGOs and others may prevent the emergence of a network. In some cases ILEIA's assistance has been requested precisely because of this competition.

Another issue is the concentration of energy and attention: should network support be broadened or deepened? At field level, competition for donor funds can sometimes make it difficult for organizations to cooperate. In these circumstances, should ILEIA put more emphasis on support to a specific group, enabling matters to move ahead relatively quickly, or seek to build contacts between groups, which may slow down the rate of progress in the short term? To what extent can we remain neutral in these situations?

On what target groups should our networking efforts focus and what are the implications? Are we placing sufficient emphasis on field-level networks? From a practical point of view, focusing on networks at the farmers' level will reduce language and travel problems, since the network's boundaries will tend to be more local. Is ILEIA in a position to link different kinds of network, or should we more explicitly opt for grass roots networks and accept the associated social, political and economic implications? Related to this question is the issue of whether ILEIA should be more involved in lobbying in the North to support changes in policies towards the South.

How to complement global and local newsletters? Some readers feel that the *ILEIA Newsletter* is becoming too abstract, containing too much jargon on LEISA (should we drop the acronym?) and not enough on practical, technical matters. Others suggest that the time seems to have passed for the *ILEIA Newsletter* to offer practical advice on, say, composting. Over the years, the newsletter has shifted its emphasis away from technical aspects to concentrate on ways of working with farmers to develop low-external-input technologies, and on socio-economic and political (policy) issues. Is this a right move?

One of the aims of ILEIA is to support local or regional information centres or libraries devoted to low-external-input and sustainable agriculture. A wealth of relevant information is now available (far more than 10 years ago), and the challenge is to make this accessible to field workers in often remote areas of developing countries. At national level this suggests a series of small information centres, each specializing in the agriculture of its own agro-ecological zone and culture (or region). Such centres could include documents on relevant local practices and methods and a list of the names and addresses of local resource persons, as well as a broader range of newsletters and other literature. On the basis of experience so far we suggest that these centres should be located within networks with a serious commitment to the promotion of low-external-input and sustainable agriculture through regional newsletters, research, training initiatives and other activities. It is also important to help organizations hosting such centres in practical library management: how to set up a library, how to manage information, how to find sources of relevant materials, how to classify documents.

Another question is how best to support the documentation of field experiences of low-external-input and sustainable agriculture. We feel that instead of supplying organizations with books from the North, the emphasis should rather be on helping people in the South to document and publish farmers' knowledge and practices. How can ILEIA best support the generation of this type of information? There is a great need for publications, but they must be adapted to local circumstances and made accessible to farmers. What kind of information should be supported if the vast majority of farmers cannot read or write? Should we place more emphasis on audio-visual communications?

Should ILEIA become involved in policy issues? Policy options for sustainable development are currently under discussion at many fora, for example within the United Nations Conference on Environment and Development (UNCED) and at the Food and Agriculture Organization (FAO) of the United Nations at the international level, and at many national conferences. To what extent should ILEIA be active at such fora? Do we have a mandate from our partners to act on their behalf, or are there more appropriate organizations to be involved in this dialogue?

In particular, should ILEIA become involved in the debate about the pricing and marketing of organic produce? Some argue that the market is the most appropriate and decentralized mechanism for decision-making on these matters. How best to reward farmers for the investment of their labour in low-external-input agriculture? Is it feasible to consider premium prices for organic produce in a developing country setting? Or will the higher prices of food produced in this way make it available only to an elite? Perhaps ILEIA can learn from the experiences of IFOAM in relation to this issue.

Lastly, we need to consider what is the appropriate future institutional arrangement for supporting networks devoted to low-external-input and sustainable agriculture in the South. What is true for its newsletter may also be true for ILEIA as a whole. Maybe the main thrust of activities is now in the South and ILEIA as an information centre based in the Netherlands should simply fade out. On the other hand, if it this were to happen it might hamper the exchange of information between North and South and make access to policy-makers, donors and researchers in the North more difficult.

Conclusion

There is great scope for regional networking in developing countries, and also great need for continuing support to the foundation and operation of such networks.

ILEIA is only one partner in the business of providing support to the networking process. We are aware that there should be room for diversity and individual approaches. We are not interested in building ILEIA networks in the South, branded with the ILEIA trademark. In addition, the aim of ILEIA is not to build new structures which duplicate or even frustrate local (in)formal structures, but rather to strengthen existing institutions in the South.

Address
Wim Hiemstra and Carine Alders, ILEIA, P.O. Box 64, 3830 AB Leusden, Netherlands.

Networking as a Development Activity: The Arid Lands Information Network

Olivia Graham

Introduction

Development projects depend on a large body of development workers who are responsible for their actual implementation. These workers are the vital link between the decision-makers and the beneficiary communities, and on their competence depends the good use of millions of pounds worth of development aid. They, in turn, depend on good information in order to replicate good practice and avoid bad practice. But their access to this information is limited by several factors: they often have fairly low levels of formal education; they often do not read or write a European language to the level required by most development literature; when they receive written information, it is often so jargon-laden and heavy that it is, in practice, inaccessible to them; they are often not considered when invitations to conferences and workshops are being handed out.

Development projects are failing, or being less efficient than they could be, simply because lessons that have been learned in one place are not being taught elsewhere. This is the context of the work of the Arid Lands Information Network (ALIN).

The Development of ALIN

Origins

ALIN was the direct result of a workshop held by Oxfam in Cotonou, Benin in 1987. The workshop was on arid lands management, and the participants were 65 Oxfam staff and partners (those whose projects were funded by Oxfam) from 12 countries in the West and East African Sahel. At the end of a week's intensive discussions on the problems encountered across the region, some of the participants expressed the view that the main usefulness of the meeting had been the opportunity to meet others doing similar work and to share experiences. The bridging of the divide between those working in anglophone and francophone Africa was felt to be especially important. Oxfam was asked to find some means of enabling this communication to continue.

The Cotonou workshop highlighted the reasons for the information vacuum, and ALIN was established to redress this situation by giving a voice to project workers and establishing a network of well informed and well motivated development workers.

Two coordinators and an administrator were recruited, based in Oxford, and left to see what could be done. It thus cannot be claimed that ALIN was originally the initiative of the field workers, except that a vague mandate was given to it at Cotonou.

At the beginning of ALIN's life, both coordinators travelled extensively in East and West Africa to find out whether the potential membership had been correctly described and what sorts of information and contact were appropriate to it. On these trips, many people expressed an interest in joining the network and were duly registered as members. The original 65 Cotonou participants, who represented the core membership, quickly grew to over 600.

The original two coordinators, although both British, had attended the Cotonou workshop and had worked for a number of years in East and West Africa. They were therefore well aware of the problem of access to information experienced by project-level development workers, and of the feelings of isolation and powerlessness that can arise when decisions are made elsewhere and the people at the bottom of the development hierarchy are left to implement them.

ALIN thus set out to:
- Identify the sort of information project-based development workers need.
- Develop ways of presenting this information that would render it most accessible to project workers.
- Provide project workers with ways in which they can raise issues, share experiences and concerns and learn from others' mistakes and setbacks.

Five years later, the ALIN Secretariat, consisting of three coordinators and three support staff, is based in Dakar and the network has more than 1200 members. It operates equally in French and English. From a recent external evaluation, the verdict of the membership is that belonging to ALIN is a positive and helpful adjunct to their work and that very few other networks or sources of information are reaching or even attempting to reach them.

In the near future, the project will be seeking the status of an independent NGO and seeking to diversify its funding by approaching other donors, although it is likely that Oxfam will continue to support it.

Network or mailing list?

Nearly 2 years into the project, the ALIN staff decided that the question of whether ALIN was a network or a mailing list ought to be tested. The difference lies in whether the members are active generators of information who can then be assisted in sharing it, or passive receivers of free information without any interest in making an effort to share it. A questionnaire was sent to each member which they had to fill in and return in order to continue their membership. The response was surprisingly high: 46% sent back completed questionnaires before the deadline, and since then many more have been received. This indication of interest and commitment by the membership gave much encouragement to the secretariat.

Much recruitment to the network is done by word of mouth, and ALIN has a policy of insisting that the individual write a letter to the ALIN office to request membership. ALIN staff have been content to let the network grow at its own rate and gain its own momentum rather than do 'cold mailings' or send out publicity to recruit members. The aim was to try to build up an active and committed membership rather than simply to establish a large mailing list. Thus it is easy to join the network, as long as a genuine interest is demonstrated.

Since the survey active participation by members has increased greatly. Their contributions to the network newsletter have made this a vehicle for the exchange of information rather than its one-way dissemination (see below). In addition, several other networking activities have developed.

Activities: Load! Fire! Aim!

The development of ALIN has been a deliberately slow and experimental journey. Robert Chambers, in his keynote address at the Cotonou workshop, suggested that it was often more productive, in development work, to try something out, see what happens and then modify it, than to come with a blueprint and try to stick to it. This is the approach of ALIN, whose staff have always seen themselves as facilitators of networking rather than providers of services. At the beginning, in the absence of any clues about how to proceed, the coordinators decided to initiate a number of activities in order to test the type of information needed, the best way of delivering it, and the uses to which it could be put.

Magazine

Baobab magazine is produced three times a year. It is edited by the ALIN staff but written mainly by network members, who use it as a platform for communication and debate. The secretariat has a policy of minimal editing of members' contributions. This is for the very important reason that a group who lack confidence in the importance of what they have to say and in their ability to say it 'correctly' may quickly become discouraged if they find their contributions being overhauled in order to make them publishable. Sometimes an article is ghost-written by an ALIN staff member (if the language of the original was very unclear); sometimes a staff member writes on the basis of conversations with ALIN members. Occasionally a 'translation' and/or synthesis of a useful but incomprehensible (to most of us!) report is done by a staff member for the magazine. There is a real challenge in presenting information accessibly to a group with low levels of literacy, but in such a way that it will not patronize the highly literate. The style of the magazine is light and lively, with much use of photos and cartoons, A4 format, large typeface and fairly tough paper. It is published in both English and French.

Over the years *Baobab* has changed slightly in style, but more so in authorship. The initial style was highly glossy, with large clear layout and much use of colour. The intention was to make it as attractive as possible to a group who were not used to reading publications. While the layout is still clear and uncomplicated, some of the glossiness has gone, and the money has gone instead into more cartoons and illustrations.

The change in authorship has been much more dramatic. In the first issue, 83% of the articles were written by Europeans; in the ninth, 81% of the articles were written by Africans.

Exchange visits

At the beginning, having little experience to draw on, the ALIN staff tried several different sorts of exchange visits. Individuals were funded to make fact-finding trips on specific subjects; development workers took the farmers they were working with to visit agricultural projects; project staff went to visit others working on similar projects; development workers took community leaders to see what had been achieved elsewhere. Visits were funded within countries, between countries and across the anglophone-francophone divide. Mistakes were made (one group simply disappeared with the money!), and undoubtedly some development tourism was funded. However, ALIN members are almost unanimously enthusiastic about these visits and are often heard to say that meeting is better than writing.

Exchange visits are seen by the ALIN staff as a way to start a process of reflection, and not an end in themselves. Reports of exchange visits are published in *Baobab* if possible, and staff try to follow up the visit later and encourage both visitors and visited to reflect on its impact.

One particularly interesting and useful visit involved four project workers from Kenya (two men and two women), who set off to find out how the Malian experience with cereal banks could help them in tackling food security problems in Kenya. They spent 3 weeks travelling around the country, visiting different climatic zones and different groups who had relevant experience. On their return they wrote an excellent report, which has been widely circulated, and had opportunities to share their new knowledge in several different fora. One of them calculated that she could cover the costs of the trip within less than a year by increasing the effectiveness and profitability of an existing stores programme.

The experience gained by the ALIN staff over 5 years and the 40 or so exchange visits funded has been analyzed and much clearer guidelines are now applied to the selection of participants, the definition of objectives, preparation for the visits and follow-up after them. For example, it has been found that a visiting group should ideally have a minimum of two people and a maximum of six; that exchanging profiles of participants with host projects is a useful preparation; and that everyone must be clear about the objectives of the visit before it takes place (this is not as simple as it sounds!). Participants are expected to work out how they will share what they have learnt with colleagues and fellow development workers in their area, and how they will write about it for other ALIN members. They are also expected to make some financial contribution to the visit if they are able to, or else their organization/project is asked to co-fund it.

Workshops and seminars

Small-scale workshops and seminars around specific themes, organized by network members, are funded by ALIN. Like visits, these are seen very much as the start of a process of reflection and discussion rather than as an end in themselves. ALIN staff do not organize and invite, but they encourage and support members to do it for themselves. Meetings tend to be kept very simple, with low organizational costs. Often the funding required is no more than the cost of food, the agenda is not pre-set but agreed by the participants at the beginning, and no per diems are paid. Transport problems are often solved by the participants.

Last year, for example, ALIN supported a small workshop at which 20 people got together to discuss the problems of grain mills. It lasted 3 days, cost very little, and proved an excellent opportunity for practical discussion. No per diems were paid, and the participants sat under a tree rather than in a stuffy meeting room. The agenda was set on the first day, and in this relaxed atmosphere everyone felt free to speak. However, the wider effect of the workshop was limited because of the poor quality of the report which was produced.

The problem of reporting of workshops has not been solved. ALIN has tried several methods—asking the participants to provide a report (what comes out is usually a bald account of the process without any of the interesting discussion); asking a journalist to attend and write it up (the journalist's view of what is interesting may not be the same as an ALIN member's); asking a participant to write an article (what comes out may be too descriptive and not analytical enough). So far, ALIN staff have resisted attending the workshops themselves, feeling that to do so would take the responsibility for the workshop away from the initiators and the participants. However, in the end this may be the best way of getting feedback.

Video

A 90-minute video (with an accompanying book) in English and French has been produced on soil and water conservation. It covers six projects in Kenya, Mali and Burkina Faso and looks at the reasons for their successes. These are then presented in terms of general development principles, such as participation, motivation, building on what is known, and so on. The video is designed to provoke thought and debate rather than to prescribe solutions. It is best used in workshops as a training tool for project workers. It has been well received in all the countries where it has been introduced, and most feedback indicates that people find it a useful tool. ALIN has been urged to produce more videos, but is unable to undertake this at the moment. The reason is that the amount of time and labour required for the first video was seriously underestimated. Even though a professional company was hired to do the filming, the complexity of making a film in two language versions and three countries, filming in both wet and dry seasons, posed serious logistical difficulties.

Booklets
Development Projects in Arid Lands is a new series of booklets on specific topics. Among the topics already covered are cereal banks, para-vet projects, credit and savings, and growing trees. These booklets come out three times a year with *Baobab* and are sent free to ALIN members. The style is similar to that of *Baobab*, with the emphasis on attractive and easy presentation using lots of pictures.

AIDS
The ALIN Secretariat sees AIDS as being a development issue of such importance that it has decided not to wait until it is raised by the membership. A series of articles on the subject will be published over the course of a year. The articles will provide basic information to network members on aspects of the disease and the epidemic, enabling them to take action if they should need to do so. These will be supplemented by a discussion pack and funding for small meetings, workshops and exchange visits where appropriate.

Meeting members
The ALIN coordinators place great importance on actually spending time with network members, to talk about their work, understand their concerns and problems and help them to communicate their expertise. For this reason they spend time and money travelling to the projects of network members. This personal contact is costly, but without it the gap between secretariat and membership could quickly become uncrossable. It is extremely useful in soliciting articles for *Baobab*, sitting with people and helping them to write down what they know and wish to share with others. The visits are also useful for discussing funding requests and work plans, and for putting people in touch with others doing similar work.

Directory of members
The Directory of ALIN Members is produced once a year and is a powerful tool for stimulating and assisting spontaneous networking between members. Great care has been taken to make it user-friendly, and each member's name and address are followed by a short paragraph describing his or her work and interests, rather than a string of incomprehensibly coded keywords. At the back is a further classification by activity, country and name, enabling, for example, someone in Burkina Faso interested in cereal banks to find out quickly who else in their country has the same interest. Up-to-date information for the directory is solicited every year from the members who wish to remain part of the network.

Issues

Where are the women?

The current membership of ALIN is 82% male and 18% female. Whether or not this reflects the actual proportion of women working in development at the project level is uncertain—statistics do not exist. But in order to raise the profile of women and gender issues, it is certainly desirable that the percentage of women increase. The percentage of articles written by women in *Baobab* is much lower—just under 10%. Of the women authors, very few have been project workers.

This concern with the proportion of women in ALIN is not cosmetic, but vital. If project-level development workers are isolated, the women among them are far more so. They are likely to be the least well informed, trained, regarded and resourced members of this already low-status group. They probably have less confidence in their judgement, knowledge and ability to communicate outside their immediate work area (although they may be powerful personalities within it). If development work does not seek to alter the existing relationship between men and women, then it cannot truly be called development. And in this, these women are the key.

But the reality is that they are very hard to reach. They have one very major constraint on their professional development—time. Working women, with domestic and family responsibilities, simply do not have the same amount of time available as men to attend meetings, workshops and training sessions, to write articles, to think about their experience and needs. If they do appear at workshops, one can be sure that getting there has been a major organizational and logistical hurdle (they have had to make alternative arrangements for the running of the household, domestic chores and childcare, and to appease husbands and other male members of the family who feel they have a call on their time).

How can we better involve women in ALIN activities from which they and their work, as well as the rest of the network, will benefit?

The first step is to encourage them to join the network. Most of the women members of ALIN have joined because they have seen someone else's copy of *Baobab* and have liked it, or they have met someone (e.g. Oxfam programme staff) who has encouraged them to write and request membership. More effort can be made to increase the use of these two channels of recruitment. Existing members are encouraged to share their copy of *Baobab*, especially with women. Certain key members who work a lot with women, for example, women's programme officers in non-government organizations (NGOs), are asked to carry copies of *Baobab* with them and to recruit women on their tours.

The fact that many women working in development are not literate in English or French can be a hurdle. But we have examples in Senegal of women who cannot read a word of French and who receive *Baobab* and get it translated out loud to them by a literate colleague in the village.

Once we have women members, the next problem is to encourage them to share their experience. This is more complicated. They will certainly not have much spare time in which to think and write. They may lack confidence that they have anything worth saying, or the ability to say it. Much encouragement is needed, and there may be several false starts, requiring sensitive handling by the ALIN Secretariat staff. Several exchanges of letters may be required after initial contact during a visit.

However, three times as many women as men have participated in exchange visits funded by ALIN, and the most successful of these have been where women have travelled together. An example was a group of 12 Burkinabé women who spent 2 weeks visiting credit and savings programmes in Togo and Benin. They were a mixed group of project animators and members of eight savings and credit cooperatives. A couple of them took their babies with them. The visit was well prepared in advance and they all travelled by road in a minibus. They were enthusiastic about what they had learned, and on their return proposed to run a series of workshops for their cooperatives, to share their experience.

Relationship between members and secretariat

The key to this issue is best thought of by considering the actual and desirable shape of the network: does it most resemble a spider's web or a bicycle wheel?

This question is a critical one for analyzing the role of the secretariat. It may be at the hub of the network, serving as the point through which most information must pass in order to reach other network members (through a magazine, newsletter or correspondence). In this role it can be very dictatorial about network activities, and entrench itself to such an extent that the network would collapse if the secretariat were to cease functioning. Or the secretariat may be one point, vaguely at the centre, of a web of networking activity which as often bypasses it as touches it. In this role, a substantial amount of spontaneous contact between members would probably be maintained if it were to cease functioning.

Most networks start off as bicycle wheels. Some never have the intention of becoming anything else. In the case of ALIN, we hope that we are gradually moving towards being a spider's web, while recognizing that, for the time being at any rate, the nature of our membership will necessitate the continuation of the secretariat in certain functions. A recent questionnaire sent out to ALIN members showed that one in four members had contacted, or been contacted by, another member as a result of an article in *Baobab*. This was an encouragingly high proportion, and feedback received so far indicates that it is far higher now that the first Directory of Members has been published.

This type of spontaneous 'horizontal' networking is difficult to monitor. ALIN has opted for an annual questionnaire which members are required to fill in and return as a condition of continued membership. The questionnaire asks them how much contact they have had with other members and whether any concrete action came out of this contact or out of reading ALIN publications. This provides an important indicator of the impact of the network.

The role of the secretariat has changed over ALIN's lifetime. At the outset the staff had a certain understanding of the communication problem between network members, and initiated *Baobab* magazine as a response to this. Since then, 'ownership' of the magazine has passed to a great extent into the hands of the members themselves through the themes they define as important and the material they contribute. However, since many of the members are very unconfident of their literary skills, the secretariat retains an important role in encouraging people to write and in working closely with them in a sympathetic and non-judgemental manner to get the best out of what they have to say. Some of this work is done when staff members travel and meet network members, some is done by correspondence. It is the work of a sympathetic editor and midwife, and is especially important when the contributor is a woman.

The secretariat staff also have a role to play in helping members who want to participate in exchange visits to clarify their objectives, to put together a fundable proposal with a realistic budget and to reflect on what they have learnt on their return. Thus when a proposal is received which is unclear and badly written, there may be a lengthy exchange of letters between the secretariat and the proposer. A proposal is never rejected before every attempt has been made to help the proposer to improve it.

A recent evaluation of ALIN suggested that one way to shift the focus of networking away from the secretariat would be to encourage the formation of country groups which could act as local foci for networking activity. This has now been tried and several experimental country groups are in existence. There are many questions to be answered as to the precise purpose of these groups. The secretariat is keen to encourage them to move away from the idea of meeting as members of ALIN and to seek an identity as like-minded people committed to networking and sharing information for better development. The role of the secretariat in this process is that of catalyst—in essence no different from the role of any development agent in helping a community to take responsibility for its own development and realize its own potential. The secretariat provides a small amount of funding for these groups, which may gradually take on a role of coordination of ALIN-funded activities, screening of proposals and channelling of funds.

Is ALIN sustainable?

For the foreseeable future, it is unlikely that ALIN would continue to be active without its secretariat. As already suggested, this has much to do with the nature of the target group. Development workers are a relatively disadvantaged group, without much confidence or influence. They will need fairly constant encouragement if they are to grow and develop in their professional lives.

This does not mean that the secretariat is not working towards a diminishing role for itself. By making other organizations (government agencies, donors) and project managers more aware of the need for better support to development workers and of the usefulness of training and information sharing, ALIN sees a reduced role for itself in the future in funding exchange visits and workshops. But most ALIN members, unlike those of networks based in the North, cannot afford to pay a membership fee which would cover the costs of running the secretariat and funding its various activities. For this, continued external funding is the only solution. However, ALIN should be seen as a long-term training and support initiative, rather than a fixed-term piece of development funding, and the issue of financial sustainability is less important than the question of whether the networking activities encouraged by ALIN are sustained.

In the long term, what ALIN staff are working towards is not greater growth in terms of size or geographical spread—a bigger club—but greater control by the members over the process of networking. In this sense, 'belonging' to ALIN does not have any meaning. It is simply a stepping stone which can be used by a presently neglected but very important section of the development community to gain greater confidence and control over their own lives and work.

Address
Olivia Graham, Arid Lands Information Network, Casier Postal 3, Dakar-Fann, Senegal.

Further Reading

Anon. 1985. Grassroots and networking: The ACFOD experience. *IFDA Dossier*, No.55:67-71.

Abeysinghe, A., Alwis, J., Medagama, J., Merry, D., Weerawardena, I.K. and Widanapathirana, A. 1991. A comprehensive comparative analysis of past and present experiences with farmer organizations in Sri Lanka. Staff Working Paper No. 23, IMPSA, Colombo, Sri Lanka.

Afonso, C.A. 1991. Telematics and the NGOs: The moment of networking. *Reflexion* 1(2):44-51.

Akhtar, S. 1990. Regional information networks: Some lessons from Latin America. *Information and Development* 6 (1):35-42.

ANEN. 1987. ANEN: The African NGOs Environmental Network. IFDA Dossier, No. 59:31-41.

Approtech Asia. 1992. Approtech Asia: Regional networking on appropriate technology and sustainable agriculture. Gate 92 (4):18-22. GTZ, Germany.

Badjagou, P.O. 1992. Efforts towards establishing a network of endogenous development organizations in the Republic of Benin. Paper presented at the International Workshop Networking for LEISA. ILEIA, Leusden, Netherlands.

Beal, G. M. and Kern, K. R. 1987. Linking knowledge systems. *Agricultural Information Development Bulletin* 9 (3):2-5.

Box, L. 1990. Agrarian knowledge networks: A conceptualisation. Sociological studies of Wageningen Agricultural University No.28:1-7. Wageningen, Netherlands.

Burley, L. 1987. International forestry research networks: Objectives, problems and management. *Unasyla* 157/158:67-73.

Burrows, S. 1989. The establishment of FAO-sponsored technical cooperation networks in Africa. RAFR Technical Papers, FAO, Rome, Italy.

Calame, P. 1993. Networks and strategies for change. IRED-Forum 46:62-67.

Caldwell, J.S. and Lightfoot, C. 1987. A network for methods of farmer-led systems experimentation. *FSSP Newsletter* 5 (4):18-24.

Chitiga-Machinguata, R.M. 1990. Networking among NGOs in Africa. *Voices from Africa* 2:65-70.

Colchester, M. 1992. Global alliance of indigenous peoples of the rainforests. *Forests, Trees and People Newsletter* 18:20-25.

Commandeur, H.R. and den Hartog, G.J. 1990. Netwerken: Terminologie, theoretische achtergonden en conceptueel kader (Networks: terminology, theoretical background and conceptual framework). Centre for Economic Research, Erasmus University, Rotterdam, Netherlands.

Compton, L.J. 1991. The role of farmer networks in minimizing risks in rainfed agriculture. Department of Continuing and Vocational Education, University of Wisconsin, USA.

COPAC. 1987. The cooperative network in developing countries: A statistical picture. Rome: FAO COPAC.

Cox, E. 1989. Networking among rural women in the Pacific. *Agricultural Information Development Bulletin* 10 (3):18-23.

de Groot, A. 1990. Cocoa-knowledge networks: The emergence of formal and informal articulation. Sociological Studies of Wageningen University No. 28:46-62. Wageningen, Netherlands.

Dodds, J.H. and Horton, D. 1989. Collaborative biotechnology networks in developing countries: The International Potato Center. *Agrotech News and Information* 1 (6): 903-906.

Durr, H.P. 1988. New paths to global cooperation: The Global Challenges Network and the International Foundation for the Survival and Development of Humanity. IFDA Dossier, No.67:51-60.

Environment Liaison Centre. 1985. Working together for sustainable development: Building NGO networks and capacities: A programme proposal. Environmental Liaison Centre, Nairobi, Kenya.

FAO 1990. Towards putting farmers in control: A second case study of the rural communication system for development in Mexico's tropical wetlands. FAO Development Communication Case Study No. 9, FAO, Rome, Italy.

Faris, D.G. 1991. Agricultural research networks as development tools: Views of a network coordinator. IDRC/ICRISAT, Ottawa, Canada/Patancheru, India.

Farrington, J. 1991. Networking for agricultural development: Fact or fashion? *Courier* 125:98-101.

Farrington, J. 1992. Synergy between research and networking at ODI: The study on 'NGOs, sustainable agricultural technology and links with the public sector' as an example. Paper presented at the International Workshop on Networking for LEISA. ILEIA, Leusden, Netherlands.

Fernandez, E.A. 1986/87. Grassroots networking: A case from Thailand. *Rural Reconstruction Review* 8:8-16.

Fernando, S. 1989. How networks function: Some structural and interactional aspects of the IRED network in Asia. Occasional Papers Series No. 3, IRED-Asia, Colombo, Sri Lanka.

Fleming, S. 1991. Between the household: Researching community organization and networks. IDS Bulletin 22 (1):37-44.

Gamage, W. 1989. An introduction to peasant exchange methodology. Series Methodologies, Strategies, and Tools No.3, IRED-Asia, Colombo, Sri Lanka.

Garcia-Padilla, V. 1992. The LEISA directory: Post-inventory observations. Paper presented at the International Workshop on Networking for LEISA. ILEIA, Leusden, Netherlands.

GRAIN. 1991. Second European Network Meeting on Genetic Resources and Biotechnology. GRAIN, Barcelona, Spain.

Further Reading

Greenland, D.J., Graswell, E.T. and Dagg, M. 1987. International networks and their potential contribution to crop and soil management research. *Outlook on Agriculture* 16 (1):42-50.

Gupta, A. 1992. The honey bee has stung. *Forests, Trees and People Newsletter* 18:8-16.

Guri, B.Y. 1992. Networking to promote ecologically sound and sustainable agriculture in Ghana: The Ecumenical Association for Sustainable Agriculture and Rural Development (ECASARD). Paper presented at the International Workshop on Networking for LEISA. ILEIA, Leusden, Netherlands.

Hariri, G. 1991. Networking in agricultural research. Paper presented at the Networking Seminar in Tunis, 24-25 April 1991. ISNAR, The Hague, Netherlands.

Herz, C. 1992. Networks: Answering a need or just a fashion? *Forests, Trees and People Newsletter* 18:26-29.

Ibikunle-Johnson, V. 1989. Operational models of the networking strategy. *Journal of the African Association for Literacy and Adult Education* 4 (2):1-5.

Ker, A.D.R. 1989. Research networks: Some IDRC experiences. Proceedings of the IBSRAM Workshop on the Management of Vertisols for Improved Agricultural Production. ICRISAT, Patancheru, India.

Killough, S. and Gonsalves, J. 1992. Networking for sustainable agriculture: The IIRR experience. Paper presented at the International Workshop on Networking for LEISA. ILEIA, Leusden, Netherlands.

Kim, S.S. 1989. Asian network on development communication by agriculture. *Journal of Korean Agricultural Education* 21 (1):17-22.

Kinney, J. 1989. National networking: The case of Malawi. *Quarterly Bulletin of the International Association of Agricultural Librarians and Documentalists* 34 (4): 180-183.

Kruiter, A. 1990. Locked networks: One-way knowledge transfer in banana production. Sociological Studies of Wageningen University No. 28:31-45. Wageningen, Netherlands.

Lipnack, J. and Stamps, J. 1986. *The networking book: People connecting with people.* Routledge and Kegan Paul, New York and London.

Lizaso, J., Johnson, S.E., Dadoun, F. and Lightfoot, C. 1991. Networking and AFSRE: A summary of discussions from the 11th Annual AFSRE Symposium, 1991. Association for Farming Systems Research and Extension, Manila, Philippines.

Lopez, V.M. 1992. The SIBAT networking strategy: Tempering through experience. Paper presented at the International Workshop on Networking for LEISA. ILEIA, Leusden, Netherlands.

Loughran, E.L. 1982. Networking, coordination, cooperation and collaboration: Different skills for different purposes. In: Brandon, J.M. and associates, Networking: A trainer's manual. Community Education Resource Center, School of Education, University of Massachusetts. Amherst, MA, USA.

Makwati, D., Keletso, K. and Rozenburg, R. 1990. Towards a national forum on sustainable agriculture: Report of the preparatory team. HIVOS, The Hague, Netherlands.

Marks, Arnaud, F. 1983. Some problems of networks: The case of CILO. *International Forum on Information and Documentation* 8 (4):17-22.

Mathen, K. 1992. Beyond networking for sustainable agriculture. Paper presented at the International Workshop on Networking for LEISA. ILEIA, Leusden, Netherlands.

McCall, M. 1987. Indigenous knowledge systems as the basis for participation: East African potentials. Working Paper No. 36, University of Twente, Enschede, Netherlands.

McCorkle, C.M., Brandsletter, R.H. and McClure, G.D. 1988. A case study on farmer innovation and communication in Niger. Academy of Educational Development, Washington D.C., USA.

Millar, D. 1992. Enhancing LEISA/PTD processes in Northern Ghana: Where we started and where we are now. Paper presented at the International Workshop on Networking for LEISA. ILEIA, Leusden, Netherlands.

Mlenge, W. 1992. Catalysts and pressure groups: Breaking barriers and building bridges between innovative practitioners. *Forests, Trees and People Newsletter* 18:17-19.

Moeliono, I. and Fisher, L. 1992. Networking for development: Some experiences and observations. Paper presented at the International Workshop on Networking for LEISA. ILEIA, Leusden, Netherlands.

Montemayor, L.Q. 1992. Networking for low-external-input and sustainable agriculture: The experience of the Free Farmers Movement. Paper presented at the International Workshop on Networking for LEISA. ILEIA, Leusden, Netherlands.

Murray, D.L. 1989. Pesticide problems and international non-governmental organizations: The Nicaraguan experience. IDA Development Anthropology Network 7 (1):6-9.

NGO Management Network. 1989. Institutional development of NGOs: An annotated bibliography. International Council of Voluntary Agencies, Geneva, Switzerland.

Owusu, D.Y. 1992. Networking to Promote LEISA: Farmers' initiatives in the Mampong Valley Social Laboratory, Ghana. Paper presented at the International Workshop on Networking for LEISA. ILEIA, Leusden, Netherlands.

Padilla, H.J. 1992. Building LEISA farmers' networks: The key to LEISA promotion. Paper presented at the International Workshop on Networking for LEISA. ILEIA, Leusden, Netherlands.

Plucknett, D.L., Smith, N.J.H. and Ozgediz, A. 1990. International agricultural research: A database of networks. CGIAR Study Paper No. 26, the World Bank, Washington D.C., USA.

Prain, G.D. and Samaniego, F.U. 1986. Beyond the farming system: On-farm commodity research in the Peruvian Highlands. Paper presented at the Farming Systems Research Symposium, Kansas, 5-8 October 1986.

Richardson, J.G. 1991. Networking: Opportunities to build effective extension programs. North Carolina State University, Raleigh, USA.

Rose, D. 1989. Database applications in tropical forestry. Serie Tecnica: Informe Tecnico No.143:16. CATIE, Turrialba, Costa Rica.

Sarason, S.B. and Lorentz, E. 1979. *The challenge of the resource exchange network.* Jossey-Bass Publishers, San Francisco, USA.

Schreckenbach, H. and Baz, P. 1992. Networking: What does it mean? Gate 92 (4):4-5.

Sison, J.C. 1990. The implementation and management of networks. *Quarterly Bulletin of the International Association of Agricultural Librarians and Documentalists* 35 (4):187-195.

Stone, M.B. 1990. The future of world agricultural information networks. *Library Trends* 8 (3):562-577.

Tandon, R. 1989. Networks as an instrument for strengthening the adult education movement. *Journal of the African Association for Literacy and Adult Education* 4 (2):13-18.

Thacker, P. 1988. Asia Pacific Women and Media Information Network. CWD Networker 1 (2 and 3):6-10.

Turton, A. 1987. Production, power and participation in rural Thailand: Experiences of poor farmers' groups. UNRISD, Geneva, Switzerland.

Vargas, V. and Lozano, I. 1991. Networking in the political proposal for building the women's movement. *Reflexion* 1 (2):33-43.

Vieta, F. 1988. Networks: Research clubs for active members. *Ceres* (FAO) 21 (3): 25-28.

Vincent, F. 1992. Networking and the strengthening of organizations: The experience of IRED. Gate 92 (4):27-29.

Vonk, R.B. 1987. The process of agroforestry extension: A case study of the CARE-Kenya Agroforestry extension project. Draft. CARE-Haiti, Port-au-Prince.

Wangola, P. 1989. Programme networks: A strategy to utilize and create free space. *Journal of the African Association for Literacy and Adult Education* 4 (2):6-12.

Wangola, P. 1989. AALAE networks: Some lessons two years later. *Journal of the African Association for Literacy and Adult Education* 4 (2):32-35.

Wassenberg, A. 1980. *Netwerken: Organisatie en strategie* (Networks: Organization and strategies). Boom Publishers, Meppel, Netherlands.

Weid, J.M. von der. 1992. Networking for LEISA: Brief outline of AS-PTA's experiences. Paper presented at the International Workshop on Networking for LEISA. ILEIA, Leusden, Netherlands.

WHO. 1986. National health development networks in support of primary health care. WHO Offset Publication No. 94, Geneva, Switzerland.

Winkelmann, D.L. 1987. Networking: Some impressions from CIMMYT. In: Webster, B., Valverde, C. and Fletcher, A. (eds), The impact of research on national agricultural development: Report on the First International Meeting of National Agricultural Research Systems and the Second IFARD Global Convention, ISNAR, The Hague, Netherlands.

Wisconsin Department of Agriculture, Trade, and Consumer Protection (DATCP). 1991. Farmer-to-farmer networks. Agricultural Resource Management Division. Madison,

WI, USA.
World Neighbors. 1992. How can networks help rural development? *World Neighbors in Action* 22:2E.
Yurjevic, A. and Altieri, M.A. 1992. CLADES: An agroecological working plan to promote sustainable rural development among resource-poor farmers in Latin America. Paper presented at the International Workshop on Networking for LEISA. ILEIA, Leusden, Netherlands.

List of Networks

Global networks

AGRECOL. Networking and Information Centre for Sustainable Agriculture in the Third World, c/o Ökozentrum, Schwengistraße 12, CH-4438 Langenbruck, Switzerland. Founded in 1983 and financed by German and Swiss government and non-government organizations, AGRECOL collects and disseminates printed and audiovisual materials on ecological smallholder farming, maintains a database on relevant projects and resource persons, advises self-help rural development groups, supports information and communication centres in the South, and operates a question-and-answer service by correspondence as well as a visitors' service in Langenbrück. It collaborates mainly with partners in Africa and the Andean regions of South America. On behalf of its African partners, AGRECOL is presently coordinating publication and distribution of *Acacia*, a newsletter on sustainable agriculture in Africa, in both French and English.

Associated Country Women of the World, 50, Warwick Square, London SW1V 2AJ, England.

Association of Farming Systems Research and Extension. Secretary, Dr. T. Finan, Bureau of Applied Research in Anthropology, University of Arizona, Tucson AZ 85721, USA. This international society promotes the development and dissemination of methods for and results of participatory on-farm systems research and extension. Such research, through the participation of both women and men farmers, encourages the development and adoption of improved and appropriate technologies and management strategies to meet the socio-economic and nutritional needs of farm families, fosters the efficient and sustainable use of natural resources, and contributes to meeting global requirements for food and fibre. Members are agricultural researchers, extension agents, development practitioners, project administrators, government planners, and donor agency representatives from more than 60 countries.

Developing Countries Farm Radio Network (DCFRN), 40 Dundas Street West, Box 12, Suite 227B, Toronto, Ontario, Canada M5G 2C2. This network produces radio scripts intended for over 100 developing countries, as well as a number of international networks. Subjects are sustainable agriculture, health and nutrition, women in farming, environment—all with a practical focus. All of the scripts are available free, in English, French or Spanish. DCFRN also issues a biannual newsletter called *Voices*. DCFRN welcomes all requests for information.

Forests, Trees and People Programme (FTPP), IRDC, Swedish University of Agricultural Sciences (SUAS), Box 7005, S-750 07 Uppsala, Sweden. This network is designed to share information in support of efforts by rural people to best utilize their natural resources, especially trees and forests. It is part of the global Forests, Trees and People Programme (FTPP), which is coordinated by the Community Forestry Unit of the FAO, Rome, Italy. The network is run as a collaborative effort between the International Rural Development Centre (IRDC) at the Swedish University of Agricultural Sciences in Uppsala, Sweden, SILVA in Paris, France and regional facilitators in Asia, Latin America and Africa. The FTPP issues the quarterly *Forests, Trees and People Newsletter*, available in English, French and Spanish, free of charge to interested institutions or individuals working in community forestry and related rural development activities. In addition, over 40 other publications related to this theme are available. There are regional contact points, a list of which is available at IRDC.

Genetic Resources Action International (GRAIN), Jonqueras 16-6-D, E-08003 Barcelona, Spain. This is a non-government, non-profit organization, which works to promote a better international order for genetic resources, based on grass roots approaches to genetic resources management, with a special focus on the contributions and needs of small-scale farmers in developing countries.

German Appropriate Technology Exchange (GATE), Sustainable Agriculture Division, c/o GTZ, Postfach 5180, D-6236 Eschborn, Germany. Founded in 1978, GATE is a special division of the German government-owned Deutsche Gesellschaft für Technische Zusammenarbeit (Germany Agency for Technical Cooperation, GTZ). ISAT (Information and Advisory Service on Appropriate Technology) is a section of GATE which collects and disseminates information on technology appropriate to the needs of developing countries. In the field of sustainable agriculture, ISAT offers a free information service, runs a documentation unit, publishes books, cooperates with non-government organizations in the South, and gives free technical consultancy to projects. Information about sustainable agriculture is also included in the quarterly *Gate bulletin* on appropriate technology.

Information Centre for Low-External-Input and Sustainable Agriculture (ILEIA), Kastanjelaan 5, P.O. Box 64, NL 3830 AB Leusden, Netherlands. Founded in 1982 and financed by the Netherlands Ministry of Development Cooperation, ILEIA documents field-level experiences in LEISA and participatory technology development (PTD). It publishes the quarterly *ILEIA Newsletter* to mobilize and share experiences worldwide in LEISA development, as well as books such as the ILEIA Readings in Sustainable Agriculture (together with Intermediate Technology Publications) and *Farming for the future: An introduction to low-external- input and sustainable agriculture* (together with Macmillan). ILEIA promotes networking among the more than 7000 individuals and groups who receive and contribute to the newsletter, supports the establishment of regional LEISA networks, and organizes workshops on key themes, such as PTD and criteria for assessing LEISA techniques.

List of Networks

International Alliance for Sustainable Agriculture (IASA), Newman Center, University of Minnesota, 1701 University Avenue SE, Minneapolis, Minnesota, 55414, USA. IASA is a not-for-profit organization, founded in 1983 to realize sustainable agriculture on a worldwide scale. It encourages networking through its newsletter, *Manna*. It has also published a directory, a resource guide and other works of interest, in English.

International Ecological Agriculture Network (IEAN). Ms. Sue Milner, 139 Rue de l'Hospice Communal, B-1170 Brussels, Belgium. As this book appears, this global network is still in its planning phase. The network aims to bring currently marginalized ecological agriculture into the mainstream and will complement the work of the numerous existing farmers' groups, networks and organizations working on agricultural development worldwide. Essentially, the IEAN will play a catalytic role in supporting, facilitating and enhancing the capacities and activities of existing networks. Network participants will include farmers' organizations, women's groups, NGOs, existing networks and policy-makers, government and donor agency representatives and research scientists.

International Federation of Organic Agricultural Movements (IFOAM), General Secretariat, c/o Ökozentrum Imsbach, D-6695 Tholey-Theley, Germany. Founded in 1972, IFOAM coordinates a network of movements around the world which promote organic farming. Its major activities involve exchanging knowledge and ideas, informing the public, representing the organic movements in international forums, holding an international scientific conference every 2 years, and setting and revising international standards for the production, processing and trade of organic products. Members include associations of producers, processors, traders and consultants, as well as institutions involved in research, training and information. IFOAM now has almost 500 full member organizations in more than 75 countries. In addition to its Internal Letter, IFOAM publishes the quarterly *Ecology and Farming*, in English with Spanish and French abstracts. Governed by a biennial General Assembly and world Board of directors, IFOAM is now decentralizing its organizational structure to better meet the needs of its member in each continent.

International Federation of Women in Agriculture. Secretary General, Dr. C. Prasad, Krishi Anusandham Bhavan, Pusa, New Delhi 110 012, India.

International Institute of Rural Reconstruction (IIRR), Silang, Cavite 4118, Philippines. This international NGO has six partner national rural reconstruction movements in various countries of the world. The IIRR became involved with sustainable agriculture in 1984. It publishes regular newsletters, and the journal *Rural Reconstruction Review*, as well as practically oriented documents, in English, on issues such as regenerative agriculture, agroforestry, upland agriculture, low-external-input rice farming, and biointensive gardening. Various training courses, on a national, regional and international level, are organized on IIRR's campus in the Philippines, on subjects such as agricultural extension, rural development and sustainable livelihoods.

Nitrogen Fixation by Tropical Agricultural Legumes (NifTAL). B. Ben Bohlool, College of Tropical Agriculture and Human Resources, Dept of Agronomy and Social Science, University of Hawai, 1000 Holomua Avenue, Paia, Hawai 96779, USA. The network's goal is to reduce the dependence of small farmers on costly nitrogen fertilizers. NifTAL seeks to reinforce research on biological nitrogen fixation in developing countries by filling gaps not covered by international and regional research centres.

Overseas Development Institute (ODI), Regent's College, Inner Circle, Regent's Park, London NW1 4NS, United Kingdom. ODI runs four networking programmes: the Agricultural Research and Extension Network (AgREN), the Irrigation Management Network (IMN), the Pastoral Development Network (PDN) and the Social Forestry Network (SFN). Communication takes place through mailings (twice a year), each including three to five substantive network papers, accompanied by a newsletter. Each network is open principally to practitioners, policy-makers and, in some cases, academics within the subject area of the network. Members are expected to send in papers or notes on their experiences.

Pesticide Action Network (PAN). PAN lobbies for better control of and information on the use of dangerous pesticides and tries to stimulate the use of biological control methods. The network has no global secretariat, but consists of numerous regional and national networks. Some examples are: PAN Indonesia, Riza V. Tjahjadi, Jl. Persada Raya 1, 12870 Jakarta, Indonesia; PAN Europe, 23 Beehive Place, Brixton, SW9 7QR London, United Kingdom; PAN North America, P.O. Box 610, San Francisco, CA 94101, USA. For more information write to PAN Germany, Gaussstraße 17, D-22765 Hamburg, Germany.

Rodale Institute, 611 Siegfriedale Road, P.O. Box 323, Kutztown, PA 19530, USA. The Rodale Institute is a non-profit educational and research organization. The institute works closely with farmers, scientists and extension agents to develop sustainable farming methods. The organization is active in worldwide networking to build up self-reliance through regenerative agriculture. Rodale organizes numerous courses on themes related to sustainability. Rodale publishes a bimonthly newsletter, *International Ag-Sieve: A Sifting of News about Regenerative Agriculture,* and also various other works of interest. Rodale Senegal (B.P. A 237, Thiès, Senegal) publishes a French version: *'Entre Nous: Bulletin d'Echange d'Informations sur l'Agriculture Régénératrice'.* The Senegalese branch also operates a question-and-answer service.

List of Networks

Sustainable Agriculture Network (SANE). M.A. Altieri, 1050 San Pablo Avenue, Albany, CA 94706, USA. At the initiative of CLADES, a proposal has been made to form a global network on sustainable agriculture, linking NGOs, universities, international and national research centres, FAO and other organizations concerned with food production and the environment and which undertake training, on-farm research and information exchange at the international, regional and local levels. An inventory of potential members has yet to be made.

World Neighbors, 4127 NW 122 Street, Oklahoma City, OK 73120-8869, USA. World Neighbors was established in 1951 as a non-profit organization with a general mandate to eliminate hunger, disease and poverty in Asia, Africa and Latin America. It tries to reach these goals through strengthening the capacity of marginalized communities to meet their basic needs. Programme priorities are food production, community-based health, family planning, water and sanitation, environmental conservation and small business. People's participation in these programmes is of paramount importance. World Neighbors produces quality learning materials, practical in nature, and offers these through a catalogue which is available on request. There is a biannual newsletter, *World Neighbors in Action*, which is published for field staff, in English, French and Spanish.

Africa

Agroforestry Research Networks for Africa (AFRENA). R.B. Scott, ICRAF, P.O. Box 30677, Nairobi, Kenya. The networks focus on multi-purpose trees for such uses as fuelwood, fodder, soil improvement, soil protection and shade. Aims are to diagnose land use problems and design agroforestry technologies to overcome these problems, develop such technologies through research, select suitable tree species and arrange training sessions.

Alley Farming Network for Tropical Africa (AFNETA), c/o IITA, Oyo Road, P.M.B. 5320, Ibadan, Nigeria. This network, established in 1989, aims to develop sustainable cropping systems based on alley farming and general agroforestry principles for various agro-ecological zones in sub-Saharan Africa. The network also aims to reduce environmental destruction by improving the efficiency and stability of land use under smallholder farming systems. The network implements activities in the area of information exchange, training and collaborative research. AFNETA publishes an English language newsletter called *Afnetan*.

Animal Traction Network for East and Southern Africa (ATNESA). Timothy E. Simalenga, Chairperson, Dept of Agricultural Engineering, Sokoine University of Agriculture, P.O. Box 3003, Morogoro, Tanzania. This regional network helps national networks to get organized and collects and collates information on organizations and individuals involved in animal traction in all countries of the region. This information is disseminated through mailing lists. A directory of people and resources is also compiled, enabling members from nearby countries to recruit each other for local consultancies. Workshops are organized and proceedings published.

Arid Lands Information Network (ALIN), C.P. 3, Dakar-Fann, Senegal. Development field staff can benefit tremendously from sharing their experiences and ideas. But network activities very often tend to reach only the more senior staff of both government and non-government organizations. Information very rarely gets to the workers at the grass roots, and their voice is seldom heard. Established in 1988, ALIN tries to facilitate networking at this grass roots level. It issues a newsletter called *Baobab* and also publishes other documents. In addition, ALIN has supported the organization of regional workshops.

Association of Church Agricultural Projects (ACDEP), P.O. Box 42, Tamale, N/R Ghana. ACDEP is a network of 24 church-based agricultural extension stations located in northern Ghana. The network aims to foster cooperation among its members, and to coordinate effective delivery of common services, such as training, information, agricultural inputs, etc. It stimulates the exchange of experiences with other regional networks and tries to provide a common voice on agricultural and rural development issues affecting the rural poor. The network runs an agricultural information centre, an inputs supply project and a family health advisor project. The network meets three times a year to discuss problems and issues of interest. Each meeting is preceded by a theme workshop.

CIMMYT Eastern and Southern Africa Economics Program (CIMMYT/ESA), P.O. Box 25171, Nairobi, Kenya. The programme's main purpose is to promote and build capacity in systems-based on-farm research techniques among national research and training institutions in Eastern and Southern Africa.

Ecumenical Association for Sustainable Agriculture and Rural Development (ECASARD). Mr. Isaac K. Darko (Secretary), P.O. Box 138, Legon, Accra, Ghana. ECASARD is a network of 22 church-based and 12 non-church-based private voluntary organizations that are all engaged in promoting agricultural development in southern Ghana. Seminars and meetings of regional representatives are organized, and a training programme has been started. Sub-committees have been set up to make recommendations for establishing a newsletter, indigenous seed collection, collection of information on indigenous veterinary practices and greater participation of women in the network.

Federation of NGOs for the Environment in Cameroon (FONGEC), c/o CIPCRE, B.P. 1256, Bafoussam, Cameroon. This network began operating in August 1991 as a forum

List of Networks

for coordination of the activities of existing environmental NGOs in Cameroon. An inventory of organizations is being made.

Forum on Sustainable Agriculture (FONSAG), Private Bag 136 Bontleng, Gaborone, Botswana. This network organizes workshops, farmers' meetings and training events to exchange experiences. The network has an extension working group and publishes the *FONSAG Newsletter*.

Natural Farming Network, c/o Zimbabwe Women's Bureau, 43 Hillside Road, Cranborne, Harare, Zimbabwe. The Natural Farming Network is an informal coalition of NGOs and Government of Zimbabwe agencies which are collaborating to arrest the impact of environmental degradation on agriculture. The network aims to create greater awareness of the need for organic agriculture and to support NGOs at grassroot level in implementing organic agriculture practices. One of the most successful activities of the network is the organization of farmer exchange visits.

Pastoral and Environmental Network in the Horn of Africa, Panther House, Room 201, West Block, 38 Mount Pleasant, London WC1X 0AP, United Kingdom.

Réseau Sénégalais d'Agriculture Durable (RESAD). Mr. Cheikh Drame, B.P. 412, Thiès, Senegal.

Réseau Développement Agriculture Durable (REDAD). Pascal Badjagou (Président), B.P. 04-0670, Cotonou, Benin. This national network consists of farmers, scientists, government organizations and NGOs. The network's objectives are to link these different kinds of people who can play a part in strengthening local initiatives for the development of sustainable agriculture; to exchange experiences; to create an information bank on innovations; and to establish contact with international organizations, so as to benefit from the experiences of other countries. The network publishes a newsletter called *Nouvelles du REDAD*.

West Africa Animal Traction Network (WAATN). Adama Faye, Département Systèmes, ISRA, B.P. 3120, Dakar, Senegal. Workshops are among the main, visible activities of this open and informal network, bringing together a multidisciplinary group of participants to exchange information. Over 140 papers have been published. There is no formal network newsletter, but documents are exchanged freely among members.

West African Farming Systems Research Network (WAFSRN), c/o IITA, P.M.B. 5320, Ibadan, Nigeria. The long-term objective of this network is to facilitate the design and implementation of a coordinated regional farming systems research programme. This objective is achieved through making inventories, training, improving the flow of information through the publication of newsletters, and organizing workshops.

Asia

Alternative Agriculture Network (AAN), c/o RRAFA (Rural Reconstruction Alumni and Friends Association), 67 Sukhumvit Soi 55, Soi Thonglore, Bangkok 10110, Thailand. This national informal network consists of over 30 NGOs, exchanging experiences in alternative agriculture. The network supports its members with training courses and research studies. It also seeks to raise public awareness of the need for alternative agriculture and to lobby at policy level.

Asian Rice Farming Systems Network (ARFSN). Dr Virgilio R. Carangal, Rice Farming Systems Program, IRRI, P.O. Box 933, Manila, Philippines. On 30 network sites all over Asia, on-farm research is carried out on topics such as cropping patterns, livestock integration, rice-fish farming and varietal testing of upland crops.

Asia-Pacific Natural Agriculture Network (APNAN), 1/4 Saint Louis 2, South Sathorn Road, Bangkok 10120, Thailand. This network of researchers was formed to study the role of effective micro-organisms in natural farming. The objective is to conduct comparable research in each participating country and to present the results in the network's own newsletter as well as in scientific journals. The network also creates training opportunities for its members and organizes conferences.

Asia Soil Conservation Network (ASOCON), P.O. Box 133 JKWB, Jakarta 10270, Indonesia. An inter-country network for programmes dealing with the problems of sustainable and environmentally sound land use at the smallholder level. The network has members from South-East Asia, including China and Papua New Guinea. ASOCON runs an information service. A data base on soil and water conservation issues is planned. ASOCON publishes a 4-monthly English language newsletter called *Contour*.

Gorakhpur Environmental Action Group (GEAG), 224 Purdilpur, M.G. College Road, P.O. Box 60, Gorakhpur 273 001, U.P. India. This group, consisting mainly of researchers and teachers concerned with sustainable agriculture, publishes a Hindi newsletter called *Vasundhara*. The mailing list has more than 200 readers and the group is trying to establish a small information centre.

Honeybee. Anil Gupta, Centre for Management in Agriculture, Indian Institute of Management, Vastrapur, Ahmedabad 380 015, India. This initiative aims to establish an international knowledge base on indigenous innovations developed by peasants and artisans. The knowledge is fed back to the farmers by means of a newsletter called

List of Networks

Honeybee, which appears in six regional editions (Hindi, Gujarati, Oriya, Tamil, Kannada and Dzongkha).

Konsorsium Pengembangan Dataran Tinggi Nusa Tenggara (The Consortium for the Development of the Nusa Tenggara Highlands), Jl. Makmur 16, Cipaganti, Bandung 40161, Indonesia. A network of NGOs and other organizations promoting sustainable agriculture in the highlands of eastern Indonesia. The central activity is a series of joint staff training events on relevant subjects. The network publishes a newsletter in Bahasa Indonesian.

Lanka Organic Agriculture Network (LOAM), c/o Ranjith de Silva, Gami Seva Sevana, Office Junction, Galaha, Sri Lanka. This network was launched at the end of 1992 following a meeting of over 200 individuals and organizations. Besides farmers/ producers, the meeting was attended by representatives from trade organizations as well as processors and exporters. LOAM hopes to improve collaboration and coordination among various organizations promoting organic agriculture in Sri Lanka. Improved marketing of organically grown products will be high on the new network's agenda, including the issue of certification.

LEISA Network, Thangameena House, Ezhil Nagar, Keeranur 622 502, Pudukkottai District, Tamil Nadu, India. This Tamil Nadu and Pondicherry regional network consists of about 60 NGOs and some 150 farmers. The network organizes courses, workshops and an on-farm trials programme. Most farmers are experimenting with LEISA technology and informal seed exchange takes place. Local informal farmer discussion groups evolve to discuss experiments. The network newsletter is called *Pasunthalir*.

Malaysian Organic Farming Network (MOFAN), c/o CETDEM, P.O. Box 382, 46740 Petaling Jaya, Malaysia.

Paguyuban Tani Hari Pangan Sedunia (World Food Day Farmers' Movement of Indonesia), Pasoran Ganjuran, P.O. Box 115, Banjul 55702, Yogyakarta, Indonesia. This farmers' organization focuses on the production as well as on the consumption side of sustainable agriculture. It organizes training events, workshops, campaigns and exchange visits.

Sibol ng Agham at Teknolohiya (SIBAT), P.O. Box 375, Central Post Office, Manila, Philippines. This network was established to revive and promote sustainable agriculture in the Philippines. The goal is to promote socio-economic development through grass roots initiatives. The network supports community initiatives in rebuilding sustainable farming systems in rural areas, through extension, training, information dissemination and community development projects. Besides other, occasional publications, two periodicals are published: the quarterly *SIBAT Newsletter* and the bi-annual *SIBOL*

Journal, both in English. SIBAT runs a regional question-and-answer service. The information centre can be visited.

Southeast Asia Sustainable Agriculture Network (SEASAN), c/o RRAFA (Rural Reconstruction Alumni and Friends Association), 67 Sukhumvit Soi 55, Soi Thonglore, Bangkok 10110, Thailand. This regional network (the Philippines, Thailand, Malaysia, Indonesia, Lao PDR, Vietnam, Cambodia and Myanmar) is a forum for the exchange of experiences on sustainable agriculture. SEASAN seeks to influence action and policy towards sustainable development and to use participatory approaches. Activities of the network are: workshops, exchange visits, newsletters, documentation and development and dissemination of appropriate education material, practical training and action research.

Southeast Asian Universities Agro-ecosystem Network (SUAN), Terd Charoenwatana, Khon Kaen University, Khon Kaen, Thailand. The network focuses on interdisciplinary human ecology research having direct relevance to policy-making for the development and management of renewable natural resources in tropical Asia.

Sustainable Agriculture Forum (Wahana Pertanian Lestari), Jl. Urip Yudomoto 108, Yogyakarta, Indonesia. This network of NGOs working in the agricultural sector was established to raise awareness of the dangers of heavy use of chemical agricultural inputs and to organize training in alternative agricultural systems.

Users' Perspective with Agricultural Research and Development (UPWARD), P.O. Box 933, Manila, Philippines. UPWARD is a network of researchers dedicated to redirecting agricultural research and development towards users, with a distinct focus on marginalized, resource-poor households. UPWARD focuses on the production and utilization of root crops. The network advocates a broad vision of food systems rather than looking at individual production factors, and supports research in the areas of production systems, genetic resources and marketing, processing and consumption. It has organized conferences and workshops at regional level, at which research findings were presented and discussed. UPWARD also sponsors several training programmes, following a train-the-trainers philosophy, publishes a newsletter in English, called *Notes from the Field*, and operates a library.

Women in Rice Farming Systems Network, c/o IRRI, P.O. Box 933, 1009 Manila, Philippines.

Latin America

Agroecología Universidad Cochabamba (AGRUCO), Fac. de Agronomía, Casilla 3392, Cochabamba, Bolivia. Launched in 1985, AGRUCO is a project of the Universidad

List of Networks

Mayor de San Simon. Its main objective is to support education and training related to ecological agriculture and indigenous knowledge within Bolivian institutions. Participatory research in Andean communities is the framework for AGRUCO's training and education activities. Facilitation and revitalization of indigenous knowledge of Andean farmers constitutes a new approach for AGRUCO. It issues a series of papers in Spanish, Serie Tecnica, and a free newsletter called *Agroecología y Saber Campesino*. It also organizes an annual international training course for university professors. AGRUCO operates an information centre, containing some 3000 documents on rural development, ecological agriculture and indigenous knowledge.

Andean Council of Ecological Management (CAME), Jr. Arequipa 128, Puño, Peru. This network consists of seven NGOs which are seeking to improve their individual work by together developing their practical knowledge and skills, organizing training and advisory services for their members. They are also seeking to influence policy-makers and public opinion.

Brazilian Agroforestry Network (REBRAF), Caixa Postal 70060, Ipanema, 22422-970 Rio de Janeiro RJ, Brazil. This network, established in 1988, promotes agroforestry alternatives to deforestation and degradation. REBRAF produces various documents on agroforestry, among which a quarterly technical newsletter, *REBRAF: Rede Brasileira Agroflorestal*. It provides specialized training courses involving field work and helps implement specific agroforestry-based development projects for low-income rural and forest dwellers' communities.

Grupo Asociado Talpuy (Talpuy Association), Apartado 222, Huancayo, Peru. Talpuy was established in the early 1980s to conduct research on the traditional practices used by farmers and to facilitate communication among farmers. This association issues a journal in Spanish, called *Minka*, every 4 months, with a distinct focus on Andean culture and identity but also with much attention to traditional sustainable agriculture. *Minka* has been the vehicle for the building up the network.

Latin American Consortium on Agroecology and Development (CLADES), Casilla 97, Correo 9, Santiago, Chile. This network of 12 Latin American NGOs was founded in 1989 to promote research, training and information exchange on eclogical agriculture and sustainable rural development in the region. The main objective is to strengthen the technical capabilities of NGO personnel so that these can promote ecologically sound, as well as culturally acceptable and economically viable, production alternatives for small-scale farmers. CLADES publishes a biannual journal, *Agroecologia y Desarollo*, which features articles and information about projects, book reviews, and analyses of contemporary rural issues.

Movimiento Campesino a Campesino, Apartado 4526, Managua, Nicaragua. Within this network, Nicaraguan peasant experimenters organize locally and nationally around peer training, food production and sustainable agriculture. Exchange visits play an important role.

Programa de Agroecologica y Desarrallo Rural (PRADER, Agroecology and Rural Development Programme), c/o SEMTA, Casilla 20410, La Paz, Bolivia. A network of 14 institutions founded in 1988.

Red de Agricultura Ecologica (RAE), c/o IDMA, Apartado Postal 110384, Lima 11, Peru. This national network on ecological agriculture consists of 50 members, nearly all NGOs. The network's activities are: coordinating members' programmes, human resource development, provision and exchange of information and influencing public opinion. The network runs a library service, organizes workshops and training and produces an information bulletin.

North America

Sustainable Farming Association (SFA) of Minnesota, P.O. Box 53, Lewiston, MN 55952, USA. This is a farmer-run educational organization dedicated to facilitating farmer-to-farmer information sharing on ecologically and economically sound farming practices. It issues a statewide newsletter called *Corner Post*.

Midwest Sustainable Agriculture Working Group (SAWG), 110 Maryland Avenue NE, Box 76, Washington D.C. 20002, USA. A network of non-profit farm, food, environmental, religious and rural organizations that advocates public policies supporting the long-term sustainability of agriculture, the conservation of natural resources and the welfare of rural communities. SAWG also works to strengthen its participating organizations through grass roots training and development programmes. The group meets quarterly.

www.ingramcontent.com/pod-product-compliance
Lightning Source LLC
Chambersburg PA
CBHW070911030426
4233 6CB000148A/2370